Everyone's Guide to Planet Uranus

Compiled by
Nicholle Rojas

Scribbles

Year of Publication 2018

ISBN : 9789352979554

Book Published by

Scribbles

(An Imprint of Alpha Editions)

email - alphaedis@gmail.com

Produced by: PediaPress GmbH
Limburg an der Lahn
Germany
http://pediapress.com/

Contents

Appendix   123

Article Licenses   135

Index   137

Introduction

Uranus

<indicator name="pp-default"> 🔒 </indicator>

Uranus

Uranus as a featureless disc, photographed by *Voyager 2* in 1986

Discovery	
Discovered by	William Herschel
Discovery date	March 13, 1781
Designations	
Pronunciation	/ˈjʊərənəs/ (🔊 listen) or /jʊəˈreɪnəs/ (🔊 listen)
Adjectives	Uranian
Orbital characteristics[1]	
Epoch J2000	
Aphelion	20.11 AU (3,008 Gm)

Perihelion	18.33 AU (2,742 Gm)
Semi-major axis	19.2184 AU (2,875.04 Gm)
Eccentricity	0.046381
Orbital period	• 84.0205 yr • 30,688.5 d • 42,718 Uranian solar days
Synodic period	369.66 days
Average orbital speed	6.80 km/s
Mean anomaly	142.238600°
Inclination	0.773° to ecliptic 6.48° to Sun's equator 1.02° to invariable plane
Longitude of ascending node	74.006°
Argument of perihelion	96.998857°
Known satellites	27
Physical characteristics	
Mean radius	25,362±7 km
Equatorial radius	25,559±4 km 4.007 Earths
Polar radius	24,973±20 km 3.929 Earths
Flattening	0.0229 ± 0.0008^2
Circumference	159,354.1 km
Surface area	8.1156×10^9 km^2 15.91 Earths
Volume	6.833×10^{13} km^3 63.086 Earths
Mass	$(8.6810\pm0.0013)\times10^{25}$ kg 14.536 Earths GM=5,793,939±13 km^3/s^2
Mean density	1.27 g/cm^3
Surface gravity	8.69 m/s^2 0.886 g
Moment of inertia factor	0.23 (estimate)

Escape velocity	21.3 km/s
Sidereal rotation period	−0.71833 d (retrograde) 17 h 14 min 24 s
Equatorial rotation velocity	2.59 km/s 9,320 km/h
Axial tilt	97.77° (to orbit)
North pole right ascension	$17^h 9^m 15^s$ 257.311°
North pole declination	−15.175°
Albedo	0.300 (Bond) 0.51 (geom.)

Surface temp.	min	mean	max
1 bar level		76 K (−197.2 °C)	
0.1 bar (tropopause)	47 K	53 K	57 K

Apparent magnitude	5.9 to 5.32
Angular diameter	3.3" to 4.1"

Atmosphere[3]

Scale height	27.7 km
Composition by volume	*(Below 1.3 bar)* **Gases**: • 83 ± 3% hydrogen (H_2) • 15 ± 3% helium (He) • 2.3% methane (CH_4) • 0.009% (0.007–0.015%) hydrogen deuteride (HD) • hydrogen sulfide (H_2S) **Ices**: • ammonia (NH_3) • water (H_2O) • ammonium hydrosulfide (NH_4SH) • methane hydrate

Uranus is the seventh planet from the Sun. It has the third-largest planetary radius and fourth-largest planetary mass in the Solar System. Uranus is similar in composition to Neptune, and both have different bulk chemical composition from that of the larger gas giants Jupiter and Saturn. For this reason, scientists often classify Uranus and Neptune as "ice giants" to distinguish them from the gas giants. Uranus's atmosphere is similar to Jupiter's and Saturn's in its primary composition of hydrogen and helium, but it contains more "ices" such as water, ammonia, and methane, along with traces of other hydrocarbons. It is the coldest planetary atmosphere in the Solar System, with a minimum temperature of 49 K (–224 °C; –371 °F), and has a complex, layered cloud structure with water thought to make up the lowest clouds and methane the uppermost layer of clouds. The interior of Uranus is mainly composed of ices and rock.

Like the other giant planets, Uranus has a ring system, a magnetosphere, and numerous moons. The Uranian system has a unique configuration among those of the planets because its axis of rotation is tilted sideways, nearly into the plane of its solar orbit. Its north and south poles, therefore, lie where most other planets have their equators. In 1986, images from *Voyager 2* showed Uranus as an almost featureless planet in visible light, without the cloud bands or storms associated with the other giant planets. Observations from Earth have shown seasonal change and increased weather activity as Uranus approached its equinox in 2007. Wind speeds can reach 250 metres per second (900 km/h; 560 mph).

Uranus is the only planet whose name is derived directly from a figure from Greek mythology, from the Latinised version of the Greek god of the sky Ouranos.

History

Like the classical planets, Uranus is visible to the naked eye, but it was never recognised as a planet by ancient observers because of its dimness and slow orbit. Sir William Herschel announced its discovery on 13 March 1781, expanding the known boundaries of the Solar System for the first time in history and making Uranus the first planet discovered with a telescope.

Discovery

Uranus had been observed on many occasions before its recognition as a planet, but it was generally mistaken for a star. Possibly the earliest known observation was by Hipparchos, who in 128 BC might have recorded it as a star for his star catalogue that was later incorporated into Ptolemy's Almagest. The earliest definite sighting was in 1690, when John Flamsteed observed it at least six times, cataloguing it as 34 Tauri. The French astronomer Pierre Charles Le Monnier observed Uranus at least twelve times between 1750 and 1769, including on four consecutive nights.

Figure 1: *William Herschel, discoverer of Uranus in 1781*

Figure 2: *Replica of the telescope used by Herschel to discover Uranus*

Sir William Herschel observed Uranus on 13 March 1781 from the garden of his house at 19 New King Street in Bath, Somerset, England (now the Herschel Museum of Astronomy), and initially reported it (on 26 April 1781) as a comet. Herschel "engaged in a series of observations on the parallax of the fixed stars",[4] using a telescope of his own design.

Herschel recorded in his journal: "In the quartile near ζ Tauri ... either [a] Nebulous star or perhaps a comet."[5] On 17 March he noted: "I looked for the Comet or Nebulous Star and found that it is a Comet, for it has changed its place."[6] When he presented his discovery to the Royal Society, he continued to assert that he had found a comet, but also implicitly compared it to a planet: <templatestyles src="Template:Quote/styles.css"/>

The power I had on when I first saw the comet was 227. From experience I know that the diameters of the fixed stars are not proportionally magnified with higher powers, as planets are; therefore I now put the powers at 460 and 932, and found that the diameter of the comet increased in proportion to the power, as it ought to be, on the supposition of its not being a fixed star, while the diameters of the stars to which I compared it were not increased in the same ratio. Moreover, the comet being magnified much beyond what its light would admit of, appeared hazy and ill-defined with these great powers, while the stars preserved that lustre and distinctness which from many thousand observations I knew they would retain. The sequel has shown that my surmises were well-founded, this proving to be the Comet we have lately observed.

Herschel notified the Astronomer Royal Nevil Maskelyne of his discovery and received this flummoxed reply from him on 23 April 1781: "I don't know what to call it. It is as likely to be a regular planet moving in an orbit nearly circular to the sun as a Comet moving in a very eccentric ellipsis. I have not yet seen any coma or tail to it."[7]

Although Herschel continued to describe his new object as a comet, other astronomers had already begun to suspect otherwise. Finnish-Swedish astronomer Anders Johan Lexell, working in Russia, was the first to compute the orbit of the new object. Its nearly circular orbit led him to a conclusion that it was a planet rather than a comet. Berlin astronomer Johann Elert Bode described Herschel's discovery as "a moving star that can be deemed a hitherto unknown planet-like object circulating beyond the orbit of Saturn".[8] Bode concluded that its near-circular orbit was more like a planet than a comet.[9]

The object was soon universally accepted as a new planet. By 1783, Herschel acknowledged this to Royal Society president Joseph Banks: "By the observation of the most eminent Astronomers in Europe it appears that the new star, which I had the honour of pointing out to them in March 1781, is a Primary Planet of our Solar System." In recognition of his achievement, King George III gave Herschel an annual stipend of £200 on condition that he move to Windsor so that the Royal Family could look through his telescopes.

Name

The name of Uranus references the ancient Greek deity of the sky Uranus (Ancient Greek: Οὐρανός), the father of Cronus (Saturn) and grandfather of Zeus (Jupiter), which in Latin became "Ūranus" (Latin pronunciation: [ˈuːranʊs]). It is the only planet whose name is derived directly from a figure of Greek mythology. The adjectival form of Uranus is "Uranian". The pronunciation of the name *Uranus* preferred among astronomers is /ˈjʊərənəs/, with stress on the first syllable as in Latin *Ūranus,* in contrast to /jʊəˈreɪnəs/, with stress on the second syllable and a long *a,* though both are considered acceptable.[10] </ref>

Consensus on the name was not reached until almost 70 years after the planet's discovery. During the original discussions following discovery, Maskelyne asked Herschel to "do the astronomical world the faver [*sic*] to give a name to your planet, which is entirely your own, [and] which we are so much obliged to you for the discovery of".[11] In response to Maskelyne's request, Herschel decided to name the object *Georgium Sidus* (George's Star), or the "Georgian Planet" in honour of his new patron, King George III. He explained this decision in a letter to Joseph Banks:

<templatestyles src="Template:Quote/styles.css"/>

> *In the fabulous ages of ancient times the appellations of Mercury, Venus, Mars, Jupiter and Saturn were given to the Planets, as being the names of their principal heroes and divinities. In the present more philosophical era it would hardly be allowable to have recourse to the same method and call it Juno, Pallas, Apollo or Minerva, for a name to our new heavenly body. The first consideration of any particular event, or remarkable incident, seems to be its chronology: if in any future age it should be asked, when this last-found Planet was discovered? It would be a very satisfactory answer to say, 'In the reign of King George the Third'.*

Herschel's proposed name was not popular outside Britain, and alternatives were soon proposed. Astronomer Jérôme Lalande proposed that it be named *Herschel* in honour of its discoverer. Swedish astronomer Erik Prosperin proposed the name *Neptune*, which was supported by other astronomers who liked the idea to commemorate the victories of the British Royal Naval fleet in the course of the American Revolutionary War by calling the new planet even *Neptune George III* or *Neptune Great Britain.*

In a March 1782 treatise, Bode proposed *Uranus*, the Latinised version of the Greek god of the sky, Ouranos.[12] Bode argued that the name should follow the mythology so as not to stand out as different from the other planets, and that Uranus was an appropriate name as the father of the first generation of the Titans. He also noted that elegance of the name in that just as Saturn was the father of Jupiter, the new planet should be named after the father of Saturn. In 1789, Bode's Royal Academy colleague Martin Klaproth named his newly discovered element uranium in support of Bode's choice. Ultimately, Bode's suggestion became the most widely used, and became universal in 1850 when HM Nautical Almanac Office, the final holdout, switched from using *Georgium Sidus* to *Uranus*.

Uranus has two astronomical symbols. The first to be proposed, ♅, was suggested by Lalande in 1784. In a letter to Herschel, Lalande described it as "un globe surmonté par la première lettre de votre nom" ("a globe surmounted by the first letter of your surname"). A later proposal, ⛢, is a hybrid of the symbols for Mars and the Sun because Uranus was the Sky in Greek mythology, which was thought to be dominated by the combined powers of the Sun and Mars.

Uranus is called by a variety of translations in other languages. In Chinese, Japanese, Korean, and Vietnamese, its name is literally translated as the "sky king star" (天王星). In Thai, its official name is *Dao Yurenat* (ดาวยูเรนัส), as in English. Its other name in Thai is *Dao Maritayu* (ดาวมฤตยู, Star of Mṛtyu), after the Sanskrit word for "death", Mrtyu (मृत्यु). In Mongolian, its name is *Tengeriin Van* (Тэнгэрийн ван), translated as "King of the Sky", reflecting its namesake god's role as the ruler of the heavens. In Hawaiian, its name is *Hele'ekala*. In Māori, its name is *Whērangi*.

Orbit and rotation

Uranus orbits the Sun once every 84 years. Its average distance from the Sun is roughly 20 AU (3 billion km; 2 billion mi). The difference between its minimum and maximum distance from the Sun is 1.8 AU, larger than that of any other planet, though not as large as that of dwarf planet Pluto.[13] The intensity of sunlight varies inversely with the square of distance, and so on Uranus (at about 20 times the distance from the Sun compared to Earth) it is about 1/400 the intensity of light on Earth. Its orbital elements were first calculated in 1783 by Pierre-Simon Laplace. With time, discrepancies began to appear between the predicted and observed orbits, and in 1841, John Couch Adams first proposed that the differences might be due to the gravitational tug of an unseen planet. In 1845, Urbain Le Verrier began his own independent research into Uranus's orbit. On 23 September 1846, Johann Gottfried Galle located a new planet, later named Neptune, at nearly the position predicted by Le Verrier.

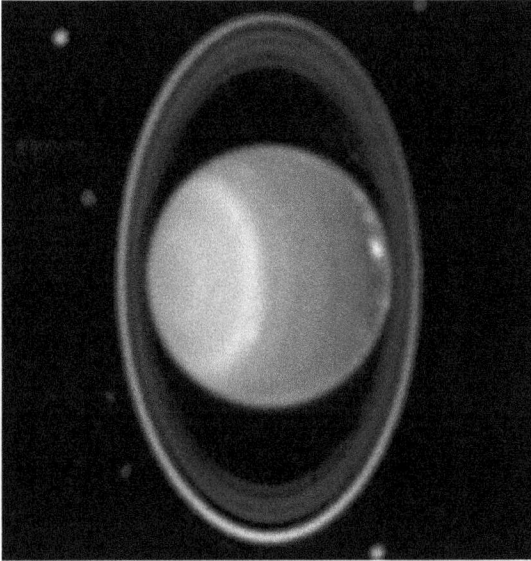

Figure 3: *A 1998 false-colour near-infrared image of Uranus showing cloud bands, rings, and moons obtained by the Hubble Space Telescope's NICMOS camera.*

The rotational period of the interior of Uranus is 17 hours, 14 minutes. As on all the giant planets, its upper atmosphere experiences strong winds in the direction of rotation. At some latitudes, such as about 60 degrees south, visible features of the atmosphere move much faster, making a full rotation in as little as 14 hours.

Axial tilt

The Uranian axis of rotation is approximately parallel with the plane of the Solar System, with an axial tilt of 97.77° (as defined by prograde rotation). This gives it seasonal changes completely unlike those of the other planets. Near the solstice, one pole faces the Sun continuously and the other faces away. Only a narrow strip around the equator experiences a rapid day–night cycle, but with the Sun low over the horizon. At the other side of Uranus's orbit the orientation of the poles towards the Sun is reversed. Each pole gets around 42 years of continuous sunlight, followed by 42 years of darkness. Near the time of the equinoxes, the Sun faces the equator of Uranus giving a period of day–night cycles similar to those seen on most of the other planets.

Uranus reached its most recent equinox on 7 December 2007.

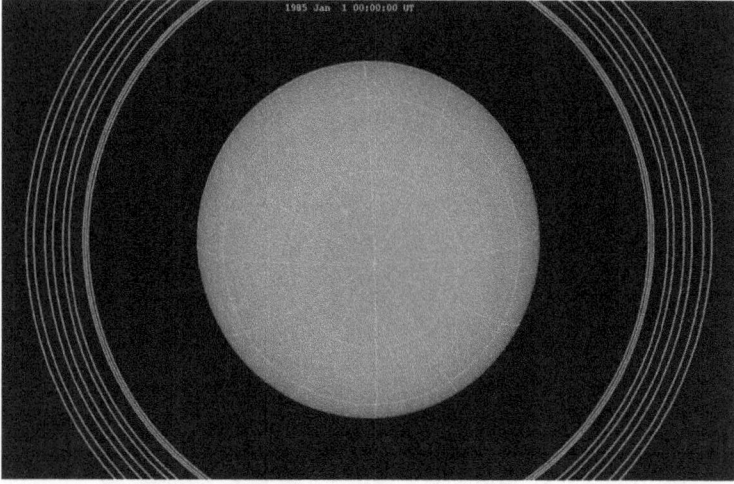

Figure 4: *Simulated Earth view of Uranus from 1986 to 2030, from southern summer solstice in 1986 to equinox in 2007 and northern summer solstice in 2028.*

Northern hemisphere	Year	Southern hemisphere
Winter solstice	1902, 1986	Summer solstice
Vernal equinox	1923, 2007	Autumnal equinox
Summer solstice	1944, 2028	Winter solstice
Autumnal equinox	1965, 2049	Vernal equinox

One result of this axis orientation is that, averaged over the Uranian year, the polar regions of Uranus receive a greater energy input from the Sun than its equatorial regions. Nevertheless, Uranus is hotter at its equator than at its poles. The underlying mechanism that causes this is unknown. The reason for Uranus's unusual axial tilt is also not known with certainty, but the usual speculation is that during the formation of the Solar System, an Earth-sized protoplanet collided with Uranus, causing the skewed orientation. Uranus's south pole was pointed almost directly at the Sun at the time of Voyager 2's flyby in 1986. The labelling of this pole as "south" uses the definition currently endorsed by the International Astronomical Union, namely that the north pole of a planet or satellite is the pole that points above the invariable plane of the Solar System, regardless of the direction the planet is spinning. A different convention is sometimes used, in which a body's north and south poles are defined according to the right-hand rule in relation to the direction of rotation.

Visibility

From 1995 to 2006, Uranus's apparent magnitude fluctuated between +5.6 and +5.9, placing it just within the limit of naked eye visibility at +6.5. Its angular diameter is between 3.4 and 3.7 arcseconds, compared with 16 to 20 arcseconds for Saturn and 32 to 45 arcseconds for Jupiter. At opposition, Uranus is visible to the naked eye in dark skies, and becomes an easy target even in urban conditions with binoculars. In larger amateur telescopes with an objective diameter of between 15 and 23 cm, Uranus appears as a pale cyan disk with distinct limb darkening. With a large telescope of 25 cm or wider, cloud patterns, as well as some of the larger satellites, such as Titania and Oberon, may be visible.

Physical characteristics

Internal structure

Uranus's mass is roughly 14.5 times that of Earth, making it the least massive of the giant planets. Its diameter is slightly larger than Neptune's at roughly four times that of Earth. A resulting density of 1.27 g/cm^3 makes Uranus the second least dense planet, after Saturn. This value indicates that it is made primarily of various ices, such as water, ammonia, and methane. The total mass of ice in Uranus's interior is not precisely known, because different figures emerge depending on the model chosen; it must be between 9.3 and 13.5 Earth masses. Hydrogen and helium constitute only a small part of the total, with between 0.5 and 1.5 Earth masses. The remainder of the non-ice mass (0.5 to 3.7 Earth masses) is accounted for by rocky material.

Figure 5: *Size comparison of Earth and Uranus*

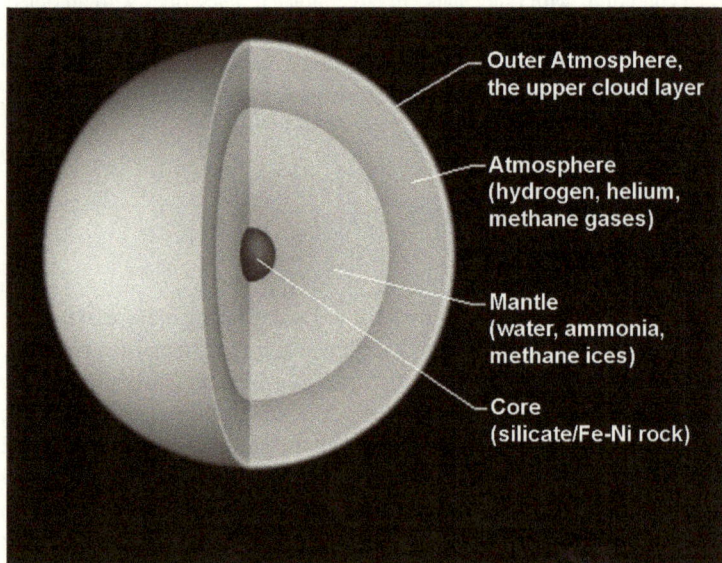

Figure 6: *Diagram of the interior of Uranus*

The standard model of Uranus's structure is that it consists of three layers: a rocky (silicate/iron–nickel) core in the centre, an icy mantle in the middle and an outer gaseous hydrogen/helium envelope. The core is relatively small, with a mass of only 0.55 Earth masses and a radius less than 20% of Uranus's; the mantle comprises its bulk, with around 13.4 Earth masses, and the upper atmosphere is relatively insubstantial, weighing about 0.5 Earth masses and extending for the last 20% of Uranus's radius. Uranus's core density is around 9 g/cm^3, with a pressure in the centre of 8 million bars (800 GPa) and a temperature of about 5000 K. The ice mantle is not in fact composed of ice in the conventional sense, but of a hot and dense fluid consisting of water, ammonia and other volatiles. This fluid, which has a high electrical conductivity, is sometimes called a water–ammonia ocean.

The extreme pressure and temperature deep within Uranus may break up the methane molecules, with the carbon atoms condensing into crystals of diamond that rain down through the mantle like hailstones. Very-high-pressure experiments at the Lawrence Livermore National Laboratory suggest that the base of the mantle may comprise an ocean of liquid diamond, with floating solid 'diamond-bergs'.

The bulk compositions of Uranus and Neptune are different from those of Jupiter and Saturn, with ice dominating over gases, hence justifying their separate classification as ice giants. There may be a layer of ionic water where the water molecules break down into a soup of hydrogen and oxygen ions, and deeper down superionic water in which the oxygen crystallises but the hydrogen ions move freely within the oxygen lattice.

Although the model considered above is reasonably standard, it is not unique; other models also satisfy observations. For instance, if substantial amounts of hydrogen and rocky material are mixed in the ice mantle, the total mass of ices in the interior will be lower, and, correspondingly, the total mass of rocks and hydrogen will be higher. Presently available data does not allow a scientific determination which model is correct. The fluid interior structure of Uranus means that it has no solid surface. The gaseous atmosphere gradually transitions into the internal liquid layers. For the sake of convenience, a revolving oblate spheroid set at the point at which atmospheric pressure equals 1 bar (100 kPa) is conditionally designated as a "surface". It has equatorial and polar radii of 25,559 ± 4 km (15,881.6 ± 2.5 mi) and 24,973 ± 20 km (15,518 ± 12 mi), respectively. This surface is used throughout this article as a zero point for altitudes.

Internal heat

Uranus's internal heat appears markedly lower than that of the other giant planets; in astronomical terms, it has a low thermal flux. Why Uranus's internal temperature is so low is still not understood. Neptune, which is Uranus's near twin in size and composition, radiates 2.61 times as much energy into space as it receives from the Sun, but Uranus radiates hardly any excess heat at all. The total power radiated by Uranus in the far infrared (i.e. heat) part of the spectrum is 1.06 ± 0.08 times the solar energy absorbed in its atmosphere. Uranus's heat flux is only 0.042 ± 0.047 W/m^2, which is lower than the internal heat flux of Earth of about 0.075 W/m^2. The lowest temperature recorded in Uranus's tropopause is 49 K (–224.2 °C; –371.5 °F), making Uranus the coldest planet in the Solar System.

One of the hypotheses for this discrepancy suggests that when Uranus was hit by a supermassive impactor, which caused it to expel most of its primordial heat, it was left with a depleted core temperature. This impact hypothesis is also used in some attempts to explain the planet's axial tilt. Another hypothesis is that some form of barrier exists in Uranus's upper layers that prevents the core's heat from reaching the surface. For example, convection may take place in a set of compositionally different layers, which may inhibit the upward heat transport; perhaps double diffusive convection is a limiting factor.

Atmosphere

Although there is no well-defined solid surface within Uranus's interior, the outermost part of Uranus's gaseous envelope that is accessible to remote sensing is called its atmosphere. Remote-sensing capability extends down to roughly 300 km below the 1 bar (100 kPa) level, with a corresponding pressure around 100 bar (10 MPa) and temperature of 320 K (47 °C; 116 °F). The tenuous thermosphere extends over two planetary radii from the nominal surface, which is defined to lie at a pressure of 1 bar. The Uranian atmosphere can be divided into three layers: the troposphere, between altitudes of –300 and 50 km (–186 and 31 mi) and pressures from 100 to 0.1 bar (10 MPa to 10 kPa); the stratosphere, spanning altitudes between 50 and 4,000 km (31 and 2,485 mi) and pressures of between 0.1 and 10^{-10} bar (10 kPa to 10 µPa); and the thermosphere extending from 4,000 km to as high as 50,000 km from the surface. There is no mesosphere.

Composition

The composition of Uranus's atmosphere is different from its bulk, consisting mainly of molecular hydrogen and helium. The helium molar fraction, i.e. the number of helium atoms per molecule of gas, is 0.15 ± 0.03 in the upper troposphere, which corresponds to a mass fraction 0.26 ± 0.05. This value is close to the protosolar helium mass fraction of 0.275 ± 0.01, indicating that helium has not settled in its centre as it has in the gas giants. The third-most-abundant component of Uranus's atmosphere is methane (CH

4). Methane has prominent absorption bands in the visible and near-infrared (IR), making Uranus aquamarine or cyan in colour. Methane molecules account for 2.3% of the atmosphere by molar fraction below the methane cloud deck at the pressure level of 1.3 bar (130 kPa); this represents about 20 to 30 times the carbon abundance found in the Sun. The mixing ratio[14] is much lower in the upper atmosphere due to its extremely low temperature, which lowers the saturation level and causes excess methane to freeze out. The abundances of less volatile compounds such as ammonia, water, and hydrogen sulfide in the deep atmosphere are poorly known. They are probably also higher than solar values. Along with methane, trace amounts of various hydrocarbons are found in the stratosphere of Uranus, which are thought to be produced from methane by photolysis induced by the solar ultraviolet (UV) radiation. They include ethane (C

2H

6), acetylene (C

2H

2), methylacetylene (CH

3C

2H), and diacetylene (C

2HC

2H). Spectroscopy has also uncovered traces of water vapour, carbon monoxide and carbon dioxide in the upper atmosphere, which can only originate from an external source such as infalling dust and comets.

Troposphere

The troposphere is the lowest and densest part of the atmosphere and is characterised by a decrease in temperature with altitude. The temperature falls from about 320 K (47 °C; 116 °F) at the base of the nominal troposphere at −300 km to 53 K (−220 °C; −364 °F) at 50 km. The temperatures in the coldest upper region of the troposphere (the tropopause) actually vary in the range between 49 and 57 K (−224 and −216 °C; −371 and −357 °F) depending on planetary latitude. The tropopause region is responsible for the vast majority of Uranus's thermal far infrared emissions, thus determining its effective temperature of 59.1 ± 0.3 K (-214.1 ± 0.3 °C; -353.3 ± 0.5 °F).

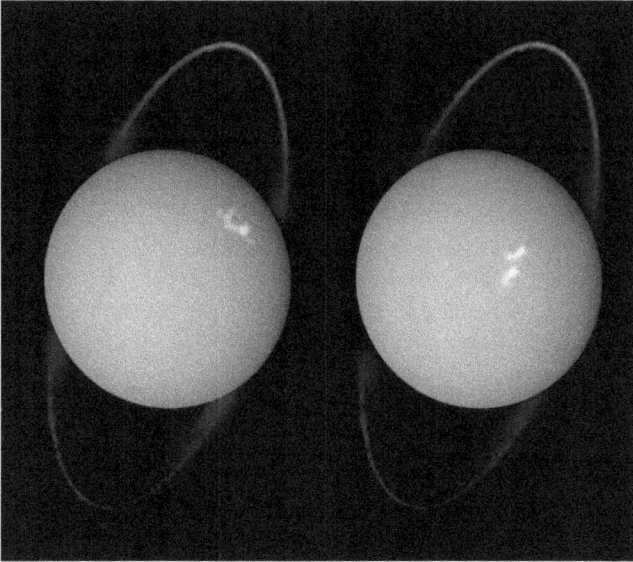

Figure 7: *Aurorae on Uranus taken by the Space Tele-scope Imaging Spectrograph (STIS) installed on Hubble.*

The troposphere is thought to have a highly complex cloud structure; water clouds are hypothesised to lie in the pressure range of 50 to 100 bar (5 to 10 MPa), ammonium hydrosulfide clouds in the range of 20 to 40 bar (2 to 4 MPa), ammonia or hydrogen sulfide clouds at between 3 and 10 bar (0.3 and 1 MPa) and finally directly detected thin methane clouds at 1 to 2 bar (0.1 to 0.2 MPa). The troposphere is a dynamic part of the atmosphere, exhibiting strong winds, bright clouds and seasonal changes.

Upper atmosphere

The middle layer of the Uranian atmosphere is the stratosphere, where temperature generally increases with altitude from 53 K (–220 °C; –364 °F) in the tropopause to between 800 and 850 K (527 and 577 °C; 980 and 1,070 °F) at the base of the thermosphere. The heating of the stratosphere is caused by absorption of solar UV and IR radiation by methane and other hydrocarbons, which form in this part of the atmosphere as a result of methane photolysis. Heat is also conducted from the hot thermosphere. The hydrocarbons occupy a relatively narrow layer at altitudes of between 100 and 300 km corresponding to a pressure range of 10 to 0.1 mbar (10.00 to 0.10 hPa) and temperatures of between 75 and 170 K (–198 and –103 °C; –325 and –154 °F). The most abundant hydrocarbons are methane, acetylene and ethane with mixing ratios of around 10^{-7} relative to hydrogen. The mixing ratio of carbon monoxide is similar at these altitudes. Heavier hydrocarbons and carbon dioxide have mixing ratios three orders of magnitude lower. The abundance ratio of water is around 7×10^{-9}. Ethane and acetylene tend to condense in the colder lower part of stratosphere and tropopause (below 10 mBar level) forming haze layers, which may be partly responsible for the bland appearance of Uranus. The concentration of hydrocarbons in the Uranian stratosphere above the haze is significantly lower than in the stratospheres of the other giant planets.

The outermost layer of the Uranian atmosphere is the thermosphere and corona, which has a uniform temperature around 800 to 850 K. The heat sources necessary to sustain such a high level are not understood, as neither the solar UV nor the auroral activity can provide the necessary energy to maintain these temperatures. The weak cooling efficiency due to the lack of hydrocarbons in the stratosphere above 0.1 mBar pressure level may contribute too. In addition to molecular hydrogen, the thermosphere-corona contains many free hydrogen atoms. Their small mass and high temperatures explain why the corona extends as far as 50,000 km (31,000 mi), or two Uranian radii, from its surface. This extended corona is a unique feature of Uranus. Its effects include a drag on small particles orbiting Uranus, causing a general depletion of dust in the Uranian rings. The Uranian thermosphere, together with the upper part of the stratosphere, corresponds to the ionosphere of Uranus. Observations show that the ionosphere occupies altitudes from 2,000 to 10,000 km (1,200 to 6,200 mi). The Uranian ionosphere is denser than that of either Saturn or Neptune, which may arise from the low concentration of hydrocarbons in the stratosphere. The ionosphere is mainly sustained by solar UV radiation and its density depends on the solar activity. Auroral activity is insignificant as compared to Jupiter and Saturn.

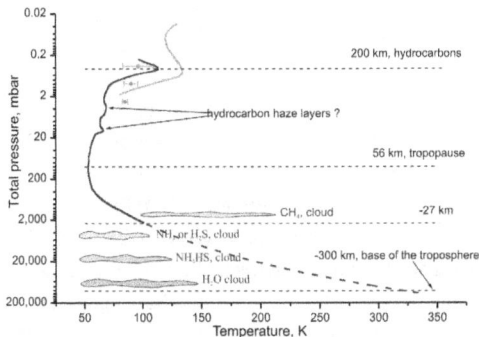

Figure 8: *Temperature profile of the Uranian troposphere and lower stratosphere. Cloud and haze layers are also indicated.*

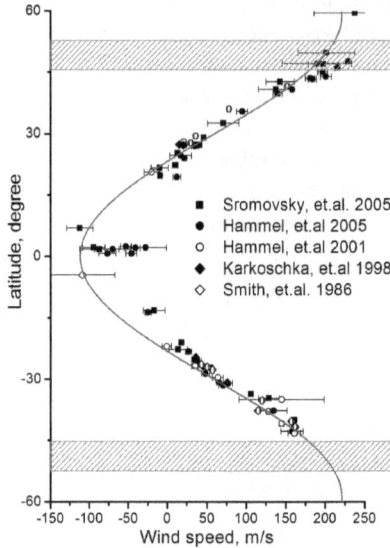

Figure 9: *Zonal wind speeds on Uranus. Shaded areas show the southern collar and its future northern counterpart. The red curve is a symmetrical fit to the data.*

Magnetosphere

Before the arrival of *Voyager 2*, no measurements of the Uranian magnetosphere had been taken, so its nature remained a mystery. Before 1986, scientists had expected the magnetic field of Uranus to be in line with the solar wind, because it would then align with Uranus's poles that lie in the ecliptic.

Voyager's observations revealed that Uranus's magnetic field is peculiar, both because it does not originate from its geometric centre, and because it is tilted at 59° from the axis of rotation. In fact the magnetic dipole is shifted from the Uranus's centre towards the south rotational pole by as much as one third of the planetary radius. This unusual geometry results in a highly asymmetric magnetosphere, where the magnetic field strength on the surface in the southern hemisphere can be as low as 0.1 gauss (10 μT), whereas in the northern hemisphere it can be as high as 1.1 gauss (110 μT). The average field at the surface is 0.23 gauss (23 μT). Studies of *Voyager 2* data in 2017 suggest that this asymmetry causes Uranus's magnetosphere to connect with the solar wind once a Uranian day, opening the planet to the Sun's particles. In comparison, the magnetic field of Earth is roughly as strong at either pole, and its "magnetic equator" is roughly parallel with its geographical equator. The dipole moment of Uranus is 50 times that of Earth. Neptune has a similarly displaced and

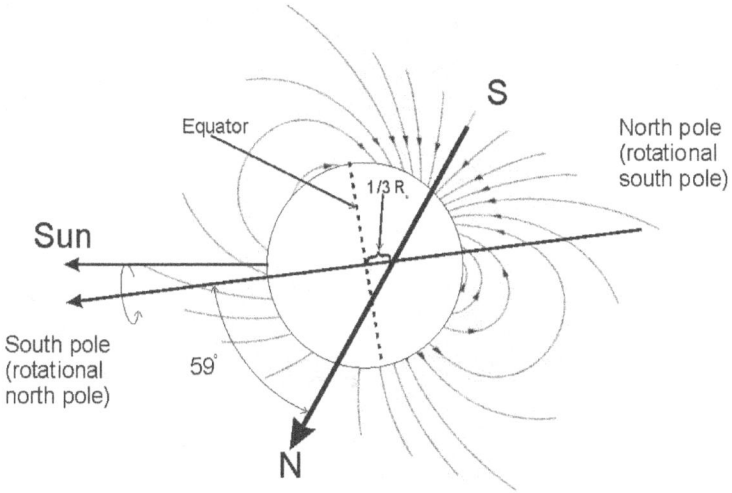

Figure 10: *The magnetic field of Uranus as observed by Voyager 2 in 1986. S and N are magnetic south and north poles.*

tilted magnetic field, suggesting that this may be a common feature of ice giants. One hypothesis is that, unlike the magnetic fields of the terrestrial and gas giants, which are generated within their cores, the ice giants' magnetic fields are generated by motion at relatively shallow depths, for instance, in the water–ammonia ocean. Another possible explanation for the magnetosphere's alignment is that there are oceans of liquid diamond in Uranus's interior that would deter the magnetic field.

Despite its curious alignment, in other respects the Uranian magnetosphere is like those of other planets: it has a bow shock at about 23 Uranian radii ahead of it, a magnetopause at 18 Uranian radii, a fully developed magnetotail, and radiation belts. Overall, the structure of Uranus's magnetosphere is different from Jupiter's and more similar to Saturn's. Uranus's magnetotail trails behind it into space for millions of kilometres and is twisted by its sideways rotation into a long corkscrew.

Uranus's magnetosphere contains charged particles: mainly protons and electrons, with a small amount of H_2^+ ions. No heavier ions have been detected. Many of these particles probably derive from the thermosphere. The ion and electron energies can be as high as 4 and 1.2 megaelectronvolts, respectively. The density of low-energy (below 1 kiloelectronvolt) ions in the inner magnetosphere is about 2 cm^{-3}. The particle population is strongly affected by the

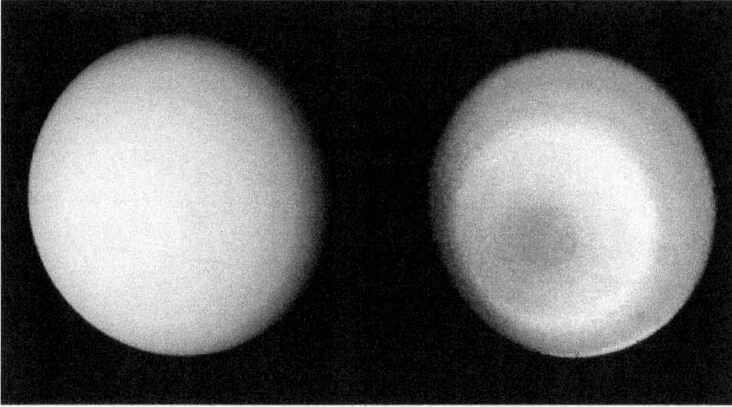

Figure 11: *Uranus's southern hemisphere in approximate nat-*
ural colour (left) and in shorter wavelengths (right), showing its
faint cloud bands and atmospheric "hood" as seen by Voyager 2

Uranian moons, which sweep through the magnetosphere, leaving noticeable gaps. The particle flux is high enough to cause darkening or space weathering of their surfaces on an astronomically rapid timescale of 100,000 years. This may be the cause of the uniformly dark colouration of the Uranian satellites and rings. Uranus has relatively well developed aurorae, which are seen as bright arcs around both magnetic poles. Unlike Jupiter's, Uranus's aurorae seem to be insignificant for the energy balance of the planetary thermosphere.

Climate

At ultraviolet and visible wavelengths, Uranus's atmosphere is bland in comparison to the other giant planets, even to Neptune, which it otherwise closely resembles. When *Voyager 2* flew by Uranus in 1986, it observed a total of ten cloud features across the entire planet. One proposed explanation for this dearth of features is that Uranus's internal heat appears markedly lower than that of the other giant planets. The lowest temperature recorded in Uranus's tropopause is 49 K (–224 °C; –371 °F), making Uranus the coldest planet in the Solar System.

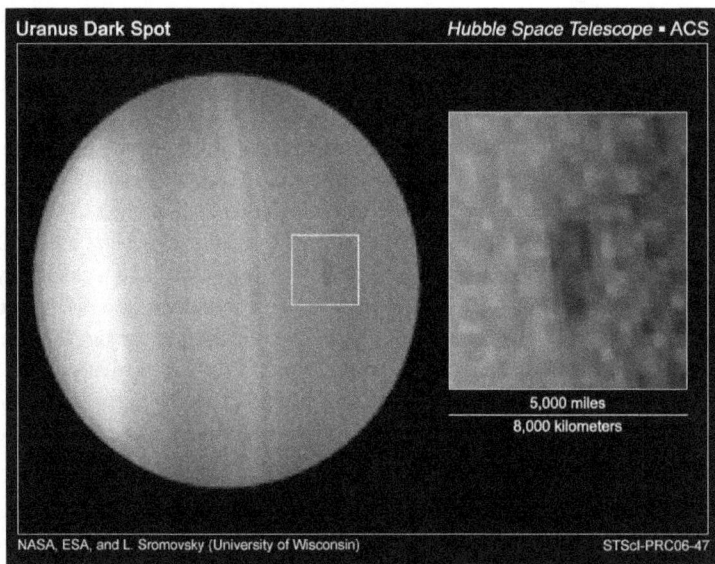

Figure 12: *The first dark spot observed on Uranus. Image obtained by the HST ACS in 2006.*

Banded structure, winds and clouds

In 1986, *Voyager 2* found that the visible southern hemisphere of Uranus can be subdivided into two regions: a bright polar cap and dark equatorial bands. Their boundary is located at about $-45°$ of latitude. A narrow band straddling the latitudinal range from -45 to $-50°$ is the brightest large feature on its visible surface. It is called a southern "collar". The cap and collar are thought to be a dense region of methane clouds located within the pressure range of 1.3 to 2 bar (see above). Besides the large-scale banded structure, Voyager 2 observed ten small bright clouds, most lying several degrees to the north from the collar. In all other respects Uranus looked like a dynamically dead planet in 1986. Voyager 2 arrived during the height of Uranus's southern summer and could not observe the northern hemisphere. At the beginning of the 21st century, when the northern polar region came into view, the Hubble Space Telescope (HST) and Keck telescope initially observed neither a collar nor a polar cap in the northern hemisphere. So Uranus appeared to be asymmetric: bright near the south pole and uniformly dark in the region north of the southern collar. In 2007, when Uranus passed its equinox, the southern collar almost disappeared, and a faint northern collar emerged near $45°$ of latitude.

In the 1990s, the number of the observed bright cloud features grew considerably partly because new high-resolution imaging techniques became available. Most were found in the northern hemisphere as it started to become visible. An early explanation—that bright clouds are easier to identify in its dark part, whereas in the southern hemisphere the bright collar masks them – was shown to be incorrect. Nevertheless there are differences between the clouds of each hemisphere. The northern clouds are smaller, sharper and brighter. They appear to lie at a higher altitude. The lifetime of clouds spans several orders of magnitude. Some small clouds live for hours; at least one southern cloud may have persisted since the *Voyager 2* flyby. Recent observation also discovered that cloud features on Uranus have a lot in common with those on Neptune. For example, the dark spots common on Neptune had never been observed on Uranus before 2006, when the first such feature dubbed Uranus Dark Spot was imaged. The speculation is that Uranus is becoming more Neptune-like during its equinoctial season.

The tracking of numerous cloud features allowed determination of zonal winds blowing in the upper troposphere of Uranus. At the equator winds are retrograde, which means that they blow in the reverse direction to the planetary rotation. Their speeds are from –360 to –180 km/h (–220 to –110 mph). Wind speeds increase with the distance from the equator, reaching zero values near ±20° latitude, where the troposphere's temperature minimum is located. Closer to the poles, the winds shift to a prograde direction, flowing with Uranus's rotation. Wind speeds continue to increase reaching maxima at ±60° latitude before falling to zero at the poles. Wind speeds at –40° latitude range from 540 to 720 km/h (340 to 450 mph). Because the collar obscures all clouds below that parallel, speeds between it and the southern pole are impossible to measure. In contrast, in the northern hemisphere maximum speeds as high as 860 km/h (540 mph) are observed near +50° latitude.

Seasonal variation

For a short period from March to May 2004, large clouds appeared in the Uranian atmosphere, giving it a Neptune-like appearance. Observations included record-breaking wind speeds of 820 km/h (510 mph) and a persistent thunderstorm referred to as "Fourth of July fireworks". On 23 August 2006, researchers at the Space Science Institute (Boulder, Colorado) and the University of Wisconsin observed a dark spot on Uranus's surface, giving scientists more insight into Uranus's atmospheric activity. Why this sudden upsurge in activity occurred is not fully known, but it appears that Uranus's extreme axial tilt results in extreme seasonal variations in its weather. Determining the nature of this seasonal variation is difficult because good data on Uranus's atmosphere have existed for less than 84 years, or one full Uranian year. Photometry over

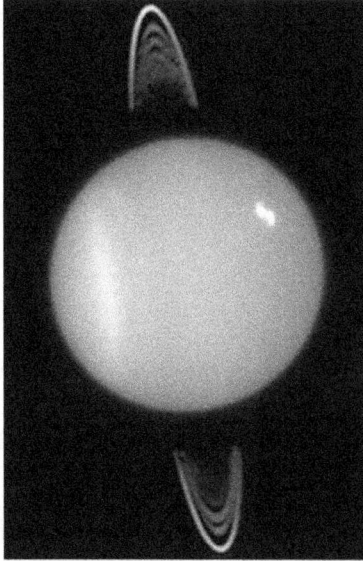

Figure 13: *Uranus in 2005. Rings, southern collar and a bright cloud in the northern hemisphere are visible (HST ACS image).*

the course of half a Uranian year (beginning in the 1950s) has shown regular variation in the brightness in two spectral bands, with maxima occurring at the solstices and minima occurring at the equinoxes. A similar periodic variation, with maxima at the solstices, has been noted in microwave measurements of the deep troposphere begun in the 1960s. Stratospheric temperature measurements beginning in the 1970s also showed maximum values near the 1986 solstice. The majority of this variability is thought to occur owing to changes in the viewing geometry.

There are some indications that physical seasonal changes are happening in Uranus. Although Uranus is known to have a bright south polar region, the north pole is fairly dim, which is incompatible with the model of the seasonal change outlined above. During its previous northern solstice in 1944, Uranus displayed elevated levels of brightness, which suggests that the north pole was not always so dim. This information implies that the visible pole brightens some time before the solstice and darkens after the equinox. Detailed analysis of the visible and microwave data revealed that the periodical changes of brightness are not completely symmetrical around the solstices, which also indicates a change in the meridional albedo patterns. In the 1990s, as Uranus moved away from its solstice, Hubble and ground-based telescopes revealed that the south polar cap darkened noticeably (except the southern collar, which remained bright), whereas the northern hemisphere demonstrated increasing activity, such as cloud formations and stronger winds, bolstering expectations that it should brighten soon. This indeed happened in 2007 when it passed an equinox: a faint northern polar collar arose, and the southern collar became nearly invisible, although the zonal wind profile remained slightly asymmetric, with northern winds being somewhat slower than southern.

The mechanism of these physical changes is still not clear. Near the summer and winter solstices, Uranus's hemispheres lie alternately either in full glare of the Sun's rays or facing deep space. The brightening of the sunlit hemisphere is thought to result from the local thickening of the methane clouds and haze layers located in the troposphere. The bright collar at −45° latitude is also connected with methane clouds. Other changes in the southern polar region can be explained by changes in the lower cloud layers. The variation of the microwave emission from Uranus is probably caused by changes in the deep tropospheric circulation, because thick polar clouds and haze may inhibit convection. Now that the spring and autumn equinoxes are arriving on Uranus, the dynamics are changing and convection can occur again.

Formation

Many argue that the differences between the ice giants and the gas giants extend to their formation. The Solar System is hypothesised to have formed from a giant rotating ball of gas and dust known as the presolar nebula. Much of the nebula's gas, primarily hydrogen and helium, formed the Sun, and the dust grains collected together to form the first protoplanets. As the planets grew, some of them eventually accreted enough matter for their gravity to hold on to the nebula's leftover gas. The more gas they held onto, the larger they became; the larger they became, the more gas they held onto until a critical point was reached, and their size began to increase exponentially. The ice giants, with only a few Earth masses of nebular gas, never reached that critical point. Recent simulations of planetary migration have suggested that both ice giants formed closer to the Sun than their present positions, and moved outwards after formation (the Nice model).

Moons

Uranus has 27 known natural satellites. The names of these satellites are chosen from characters in the works of Shakespeare and Alexander Pope. The five main satellites are Miranda, Ariel, Umbriel, Titania, and Oberon. The Uranian satellite system is the least massive among those of the giant planets; the combined mass of the five major satellites would be less than half that

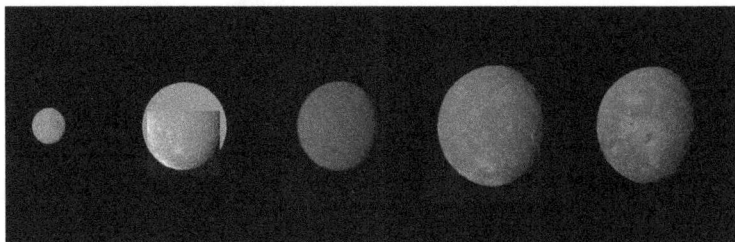

Figure 14: *Major moons of Uranus in order of increasing distance (left to right), at their proper relative sizes and albedos (collage of Voyager 2 photographs)*

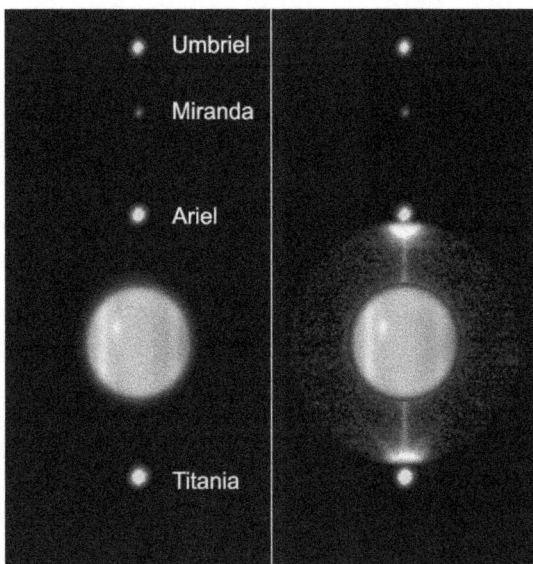

Figure 15: *The Uranus System (NACO/VLT image)*

of Triton (largest moon of Neptune) alone. The largest of Uranus's satellites, Titania, has a radius of only 788.9 km (490.2 mi), or less than half that of the Moon, but slightly more than Rhea, the second-largest satellite of Saturn, making Titania the eighth-largest moon in the Solar System. Uranus's satellites have relatively low albedos; ranging from 0.20 for Umbriel to 0.35 for Ariel (in green light). They are ice–rock conglomerates composed of roughly 50% ice and 50% rock. The ice may include ammonia and carbon dioxide.

Among the Uranian satellites, Ariel appears to have the youngest surface with the fewest impact craters and Umbriel's the oldest. Miranda has fault canyons

20 km (12 mi) deep, terraced layers, and a chaotic variation in surface ages and features. Miranda's past geologic activity is thought to have been driven by tidal heating at a time when its orbit was more eccentric than currently, probably as a result of a former 3:1 orbital resonance with Umbriel. Extensional processes associated with upwelling diapirs are the likely origin of Miranda's 'racetrack'-like coronae. Ariel is thought to have once been held in a 4:1 resonance with Titania.

Uranus has at least one horseshoe orbiter occupying the Sun–Uranus L_3 Lagrangian point—a gravitationally unstable region at 180° in its orbit, 83982 Crantor. Crantor moves inside Uranus's co-orbital region on a complex, temporary horseshoe orbit. 2010 EU$_{65}$ is also a promising Uranus horseshoe librator candidate.

Planetary rings

The Uranian rings are composed of extremely dark particles, which vary in size from micrometres to a fraction of a metre. Thirteen distinct rings are presently known, the brightest being the ε ring. All except two rings of Uranus are extremely narrow – they are usually a few kilometres wide. The rings are probably quite young; the dynamics considerations indicate that they did not form with Uranus. The matter in the rings may once have been part of a moon (or moons) that was shattered by high-speed impacts. From numerous pieces of debris that formed as a result of those impacts, only a few particles survived, in stable zones corresponding to the locations of the present rings.

William Herschel described a possible ring around Uranus in 1789. This sighting is generally considered doubtful, because the rings are quite faint, and in the two following centuries none were noted by other observers. Still, Herschel made an accurate description of the epsilon ring's size, its angle relative to Earth, its red colour, and its apparent changes as Uranus travelled around the Sun. The ring system was definitively discovered on 10 March 1977 by James L. Elliot, Edward W. Dunham, and Jessica Mink using the Kuiper Airborne Observatory. The discovery was serendipitous; they planned to use the occultation of the star SAO 158687 (also known as HD 128598) by Uranus to study its atmosphere. When their observations were analysed, they found that the star had disappeared briefly from view five times both before and after it disappeared behind Uranus. They concluded that there must be a ring system around Uranus. Later they detected four additional rings. The rings were directly imaged when Voyager 2 passed Uranus in 1986. Voyager 2 also discovered two additional faint rings, bringing the total number to eleven.

In December 2005, the Hubble Space Telescope detected a pair of previously unknown rings. The largest is located twice as far from Uranus as the previously known rings. These new rings are so far from Uranus that they are called the "outer" ring system. Hubble also spotted two small satellites, one of which, Mab, shares its orbit with the outermost newly discovered ring. The new rings bring the total number of Uranian rings to 13. In April 2006, images of the new rings from the Keck Observatory yielded the colours of the outer rings: the outermost is blue and the other one red. One hypothesis concerning the outer ring's blue colour is that it is composed of minute particles of water ice from the surface of Mab that are small enough to scatter blue light. In contrast, Uranus's inner rings appear grey.

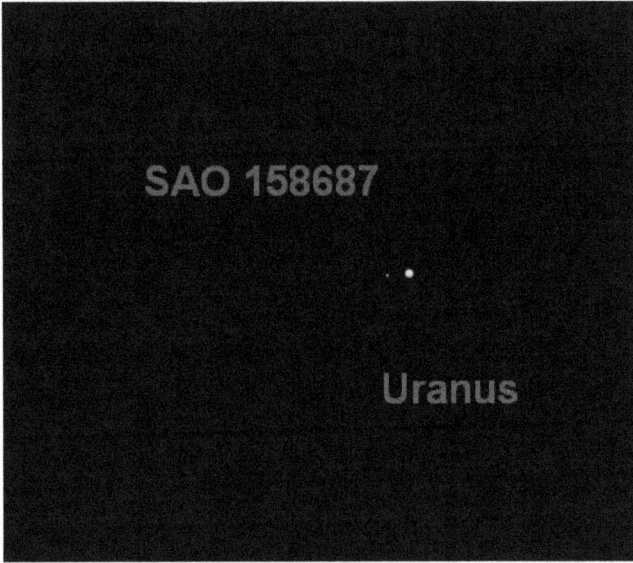

Figure 16: *Animation about the discovering occultation in 1977. (Click on it to start)*

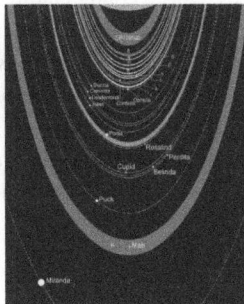

Figure 17: *Uranus has a complicated planetary ring system, which was the second such system to be discovered in the Solar System after Saturn's. UNIQ-ref-1-cf064b24719039b5-QINU*

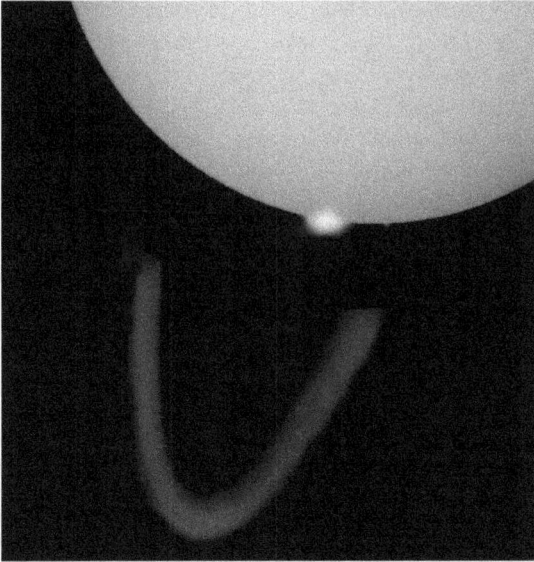

Figure 18: *Uranus's aurorae against its equatorial rings, imaged by the Hubble telescope. Unlike the aurorae of Earth and Jupiter, those of Uranus are not in line with its poles, due to its lopsided magnetic field.*

Exploration

In 1986, NASA's *Voyager 2* interplanetary probe encountered Uranus. This flyby remains the only investigation of Uranus carried out from a short distance and no other visits are planned. Launched in 1977, *Voyager 2* made its closest approach to Uranus on 24 January 1986, coming within 81,500 km (50,600 mi) of the cloudtops, before continuing its journey to Neptune. The spacecraft studied the structure and chemical composition of Uranus's atmosphere, including its unique weather, caused by its axial tilt of 97.77°. It made the first detailed investigations of its five largest moons and discovered 10 new ones. It examined all nine of the system's known rings and discovered two more. It also studied the magnetic field, its irregular structure, its tilt and its unique corkscrew magnetotail caused by Uranus's sideways orientation.

Voyager 1 was unable to visit Uranus because investigation of Saturn's moon Titan was considered a priority. This trajectory took *Voyager 1* out of the plane of the ecliptic, ending its planetary science mission.[118]

The possibility of sending the *Cassini* spacecraft from Saturn to Uranus was evaluated during a mission extension planning phase in 2009, but was ultimately rejected in favour of destroying it in the Saturnian atmosphere. It would

Figure 19: *Crescent Uranus as imaged by Voyager 2 while en route to Neptune*

have taken about twenty years to get to the Uranian system after departing Saturn. A Uranus orbiter and probe was recommended by the 2013–2022 Planetary Science Decadal Survey published in 2011; the proposal envisages launch during 2020–2023 and a 13-year cruise to Uranus. A Uranus entry probe could use Pioneer Venus Multiprobe heritage and descend to 1–5 atmospheres. The ESA evaluated a "medium-class" mission called Uranus Pathfinder.[15] A New Frontiers Uranus Orbiter has been evaluated and recommended in the study, *The Case for a Uranus Orbiter*. Such a mission is aided by the ease with which a relatively big mass can be sent to the system—over 1500 kg with an Atlas 521 and 12-year journey. For more concepts see Proposed Uranus missions.

In culture

In astrology, the planet Uranus (♅) is the ruling planet of Aquarius. Because Uranus is cyan and Uranus is associated with electricity, the colour electric blue, which is close to cyan, is associated with the sign Aquarius (see Uranus in astrology).

The chemical element uranium, discovered in 1789 by the German chemist Martin Heinrich Klaproth, was named after the newly discovered planet Uranus.

"Uranus, the Magician" is a movement in Gustav Holst's orchestral suite *The Planets*, written between 1914 and 1916.

Operation Uranus was the successful military operation in World War II by the Soviet army to take back Stalingrad and marked the turning point in the land war against the Wehrmacht.

The lines "Then felt I like some watcher of the skies/When a new planet swims into his ken", from John Keats's "On First Looking Into Chapman's Homer", are a reference to Herschel's discovery of Uranus.

Many references to Uranus in popular culture and news involve humor about one pronunciation of its name resembling that of the phrase "your anus".

Further reading

- Alexander, Arthur Francis O'Donel (1965). *The Planet Uranus – A History of Observation, Theory and Discovery.*
- Miner, Ellis D. (1998). *Uranus: The Planet, Rings and Satellites.* New York: John Wiley and Sons. ISBN 978-0-471-97398-0.
- Bode, Johann Elert (1784). *Von dem neu entdeckten Planeten*[16]. Verfasser.
- Gore, Rick (August 1986). "Uranus — Voyager Visits a Dark Planet". *National Geographic.* Vol. 170 no. 2. pp. 178–194. ISSN 0027-9358[17]. OCLC 643483454[18].

External links

●)))	Wikiquote has quotations related to: *Uranus*

- Uranus[19] at *Encyclopædia Britannica*
- Uranus[20] at European Space Agency
- NASA's Uranus fact sheet[21]
- Uranus Profile[22] at NASA's Solar System Exploration site[23]
- Planets – Uranus[24] A kid's guide to Uranus.
- Uranus[25] at Jet Propulsion Laboratory's planetary photojournal. (photos)
- Voyager at Uranus[26] (photos)
- Uranus (Astronomy Cast homepage)[27] (blog)
- Uranian system montage[28] (photo)
- Gray, Meghan; Merrifield, Michael (2010). "Uranus"[29]. *Sixty Symbols.* Brady Haran for the University of Nottingham.
- "How to Pronounce Uranus"[30] by CGP Grey

<indicator name="featured-star"> ⭐ </indicator>

Atmosphere

Atmosphere of Uranus

<indicator name="good-star"> ⊕ </indicator>

The **atmosphere of Uranus** is composed primarily of hydrogen and helium. At depth it is significantly enriched in volatiles (dubbed "ices") such as water, ammonia and methane. The opposite is true for the upper atmosphere, which contains very few gases heavier than hydrogen and helium due to its low temperature. Uranus's atmosphere is the coldest of all the planets, with its temperature reaching as low as 49 K.

The Uranian atmosphere can be divided into three main layers: the troposphere, between altitudes of –300 and 50 km and pressures from 100 to 0.1 bar; the stratosphere, spanning altitudes between 50 and 4000 km and pressures of between 0.1 and 10^{-10} bar; and the hot thermosphere (and exosphere) extending from an altitude of 4,000 km to several Uranian radii from the nominal surface at 1 bar pressure.[31] Unlike Earth's, Uranus's atmosphere has no mesosphere.

The troposphere hosts four cloud layers: methane clouds at about 1.2 bar, hydrogen sulfide and ammonia clouds at 3–10 bar, ammonium hydrosulfide clouds at 20–40 bar, and finally water clouds below 50 bar. Only the upper two cloud layers have been observed directly—the deeper clouds remain speculative. Above the clouds lie several tenuous layers of photochemical haze. Discrete bright tropospheric clouds are rare on Uranus, probably due to sluggish convection in the planet's interior. Nevertheless, observations of such clouds were used to measure the planet's zonal winds, which are remarkably fast with speeds up to 240 m/s.

Little is known about the Uranian atmosphere as to date only one spacecraft, *Voyager 2*, which passed by the planet in 1986, obtained some valuable compositional data. No other missions to Uranus are currently scheduled.

Figure 20: *Uranus by Voyager 2*

Observation and exploration

Although there is no well-defined solid surface within Uranus's interior, the outermost part of Uranus's gaseous envelope (the region accessible to remote sensing) is called its atmosphere.[31] Remote sensing capability extends down to roughly 300 km below the 1 bar level, with a corresponding pressure around 100 bar and temperature of 320 K.[32]

The observational history of the Uranian atmosphere is long and full of error and frustration. Uranus is a relatively faint object, and its visible angular diameter is smaller than 4".[33] The first spectra of Uranus were observed through a prism in 1869 and 1871 by Angelo Secchi and William Huggins, who found a number of broad dark bands, which they were unable to identify.[33] They also failed to detect any solar Fraunhofer lines—the fact later interpreted by Norman Lockyer as indicating that Uranus emitted its own light as opposed to reflecting light from the Sun.[33,34] In 1889 however, astronomers observed solar Fraunhofer lines in photographic ultraviolet spectra of the planet, proving once and for all that Uranus was shining by reflected light.[35] The nature of the broad dark bands in its visible spectrum remained unknown until the fourth decade of the twentieth century.[33]

The key to deciphering Uranus's spectrum was found in the 1930s by Rupert Wildt and Vesto Slipher,[36] who found that the dark bands at 543, 619, 925,

865 and 890 nm belonged to gaseous methane.[33] They had never been observed before because they were very weak and required a long path length to be detected.[36] This meant that the atmosphere of Uranus was transparent to a much greater depth compared to those of other giant planets.[33] In 1950, Gerard Kuiper noticed another diffuse dark band in the spectrum of Uranus at 827 nm, which he failed to identify.[37] In 1952 Gerhard Herzberg, a future Nobel Prize winner, showed that this band was caused by the weak quadrupole absorption of molecular hydrogen, which thus became the second compound detected on Uranus.[38] Until 1986 only two gases, methane and hydrogen, were known in the Uranian atmosphere.[33] The far-infrared spectroscopic observation beginning from 1967 consistently showed the atmosphere of Uranus was in approximate thermal balance with incoming solar radiation (in other words, it radiated as much heat as it received from the Sun), and no internal heat source was required to explain observed temperatures.[39] No discrete features had been observed on Uranus prior to the *Voyager 2* visit in 1986.[40]

In January 1986, the *Voyager 2* spacecraft flew by Uranus at a minimal distance of 107,100 km[41] providing the first close-up images and spectra of its atmosphere. They generally confirmed that the atmosphere was made of mainly hydrogen and helium with around 2% methane.[42] The atmosphere appeared highly transparent and lacking thick stratospheric and tropospheric hazes. Only a limited number of discrete clouds were observed.[43]

In the 1990s and 2000s, observations by the Hubble Space Telescope and by ground-based telescopes equipped with adaptive optics systems (the Keck telescope and NASA Infrared Telescope Facility, for instance) made it possible for the first time to observe discrete cloud features from Earth.[44] Tracking them has allowed astronomers to re-measure wind speeds on Uranus, known before only from the *Voyager 2* observations, and to study the dynamics of the Uranian atmosphere.[45]

Composition

The composition of the Uranian atmosphere is different from that of Uranus as a whole, consisting mainly of molecular hydrogen and helium.[46] The helium molar fraction, i.e. the number of helium atoms per molecule of hydrogen/helium, was determined from the analysis of *Voyager 2* far infrared and radio occultation observations.[47] The currently accepted value is 0.152 ± 0.033 in the upper troposphere, which corresponds to a mass fraction 0.262 ± 0.048.[46,48] This value is very close to the protosolar helium mass fraction of 0.2741 ± 0.0120,[49] indicating that helium has not settled towards the centre of the planet as it has in the gas giants.[50]

The third most abundant constituent of the Uranian atmosphere is methane (CH_4),[51] the presence of which has been known for some time as a result of the ground-based spectroscopic observations.[46] Methane possesses prominent absorption bands in the visible and near-infrared, making Uranus aquamarine or cyan in colour.[52] Below the methane cloud deck at 1.3 bar methane molecules account for about 2.3%[53] of the atmosphere by molar fraction; about 10 to 30 times that found in the Sun.[46,47] The mixing ratio is much lower in the upper atmosphere due to the extremely low temperature at the tropopause, which lowers the saturation level and causes excess methane to freeze out.[54] Methane appears to be undersaturated in the upper troposphere above the clouds having a partial pressure of only 30% of the saturated vapor pressure there.[53] The concentration of less volatile compounds such as ammonia, water and hydrogen sulfide in the deep atmosphere is poorly known.[46] However, as with methane, their abundances are probably greater than solar values by a factor of at least 20 to 30,[55] and possibly by a factor of a few hundred.[56]

Knowledge of the isotopic composition of Uranus's atmosphere is very limited.[57] To date the only known isotope abundance ratio is that of deuterium to light hydrogen: 5.5+3.5 –1.5×10^{-5}, which was measured by the Infrared Space Observatory (ISO) in the 1990s. It appears to be higher than the protosolar value of (2.25±0.35)×10^{-5} measured in Jupiter.[58] The deuterium is found almost exclusively in hydrogen deuteride molecules which it forms with normal hydrogen atoms.[59]

Infrared spectroscopy, including measurements with Spitzer Space Telescope (SST),[60] and UV occultation observations,[61] found trace amounts of complex hydrocarbons in the stratosphere of Uranus, which are thought to be produced from methane by photolysis induced by solar UV radiation.[62] They include ethane (C_2H_6), acetylene (C_2H_2),[61,63] methylacetylene (CH_3C_2H), diacetylene (C_2HC_2H).[64] Infrared spectroscopy also uncovered traces of water vapour,[65] carbon monoxide[66] and carbon dioxide in the stratosphere, which are likely to originate from an external source such as infalling dust and comets.[64]

Structure

The Uranian atmosphere can be divided into three main layers: the troposphere, between altitudes of –300[67] and 50 km and pressures from 100 to 0.1 bar; the stratosphere, spanning altitudes between 50 and 4000 km and pressures between 0.1 and 10^{-10} bar; and the thermosphere/exosphere extending from 4000 km to as high as a few Uranus radii from the surface. There is no mesosphere.[31,68]

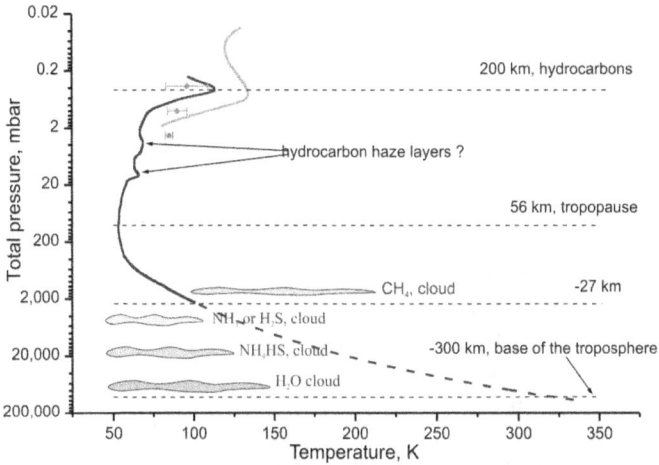

Figure 21: *Temperature profile of the Uranian troposphere and lower stratosphere. Cloud and haze layers are also indicated.*

Troposphere

The troposphere is the lowest and densest part of the atmosphere and is characterised by a decrease in temperature with altitude.[31] The temperature falls from about 320 K at the base of the troposphere at –300 km to about 53 K at 50 km.[32,47] The temperature at the cold upper boundary of the troposphere (the tropopause) actually varies in the range between 49 and 57 K depending on planetary latitude, with the lowest temperature reached near 25° southern latitude.[69,70] The troposphere holds almost all of the mass of the atmosphere, and the tropopause region is also responsible for the vast majority of the planet's thermal far infrared emissions, thus determining its effective temperature of 59.1±0.3 K.[70,71]

The troposphere is believed to possess a highly complex cloud structure; water clouds are hypothesised to lie in the pressure range of 50 to 300 bar, ammonium hydrosulfide clouds in the range of 20 and 40 bar, ammonia or hydrogen sulfide clouds at between 3 and 10 bar and finally thin methane clouds at 1 to 2 bar.[32,52,55] Although *Voyager 2* directly detected methane clouds,[53] all other cloud layers remain speculative. The existence of a hydrogen sulfide cloud layer is only possible if the ratio of sulfur and nitrogen abundances (S/N ratio) is significantly greater than its solar value of 0.16.[52] Otherwise all hydrogen sulfide would react with ammonia, producing ammonium hydrosulfide, and the ammonia clouds would appear instead in the pressure range 3–10 bar.[56] The elevated S/N ratio implies depletion of ammonia in the pressure range 20–40 bar, where the ammonium hydrosulfide clouds form. These can result from the dissolution of ammonia in water droplets within water clouds or in the deep water-ammonia ionic ocean.[55,56]

The exact location of the upper two cloud layers is somewhat controversial. Methane clouds were directly detected by *Voyager 2* at 1.2–1.3 bar by radio occultation.[53] This result was later confirmed by an analysis of the *Voyager 2* limb images.[52] The top of the deeper ammonia/hydrogen sulfide clouds were determined to be at 3 bar based on the spectroscopic data in the visible and near-infra spectral ranges (0.5–1 μm).[72] However a recent analysis of the spectroscopic data in the wavelength range 1–2.3 μm placed the methane cloudtops at 2 bar, and the top of the lower clouds at 6 bar.[73] This contradiction may be resolved when new data on methane absorption in Uranus's atmosphere are available.[74,75] </ref> The optical depth of the two upper cloud layers varies with latitude: both become thinner at the poles as compared to the equator, though in 2007 the methane cloud layer's optical depth had a local maximum at 45°S, where the southern polar collar is located (see below).[76]

The troposphere is very dynamic, exhibiting strong zonal winds, bright methane clouds,[77] dark spots[78] and noticeable seasonal changes. (see below)[79]

Stratosphere

The stratosphere is the middle layer of the Uranian atmosphere, in which temperature generally increases with altitude from 53 K in the tropopause to between 800 and 850 K at the base thermosphere.[80] The heating of the stratosphere is caused by the downward heat conduction from the hot thermosphere[81,82] as well as by absorption of solar UV and IR radiation by methane and the complex hydrocarbons formed as a result of methane photolysis.[62,81] The methane enters the stratosphere through the cold tropopause, where its mixing ratio relative to molecular hydrogen is about 3×10^{-5}, three times below saturation.[54] It decreases further to about 10^{-7} at the altitude corresponding to pressure of 0.1 mbar.[83]

Hydrocarbons heavier than methane are present in a relatively narrow layer between 160 and 320 km in altitude, corresponding to the pressure range from 10 to 0.1 mbar and temperatures from 100 to 130 K.[54,64] The most abundant stratospheric hydrocarbons after methane are acetylene and ethane, with mixing ratios of around 10^{-7}.[83] Heavier hydrocarbons like methylacetylene and diacetylene have mixing ratios of about 10^{-10}—three orders of magnitude lower.[64] The temperature and hydrocarbon mixing ratios in the stratosphere vary with time and latitude.[84,85] at the poles the hydrocarbons were also confined to much lower altitudes.[86] Temperatures in the stratosphere may increase

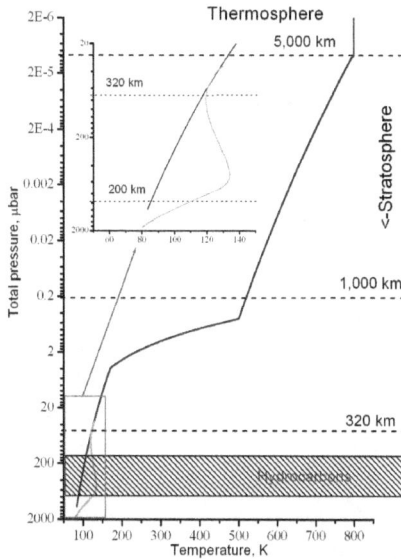

Figure 22: *Temperature profiles in the stratosphere and thermosphere of Uranus. The shaded area is where hydrocarbons are concentrated.*

at the solstices and decrease at equinoxes by as much as 50 K.[87] </ref> Complex hydrocarbons are responsible for the cooling of the stratosphere, especially acetylene, having a strong emission line at the wavelength of 13.7 μm.[81]

In addition to hydrocarbons, the stratosphere contains carbon monoxide, as well as traces of water vapor and carbon dioxide. The mixing ratio of carbon monoxide—3 × 10^{-8}—is very similar to that of the hydrocarbons,[66] while the mixing ratios of carbon dioxide and water are about 10^{-11} and 8 × 10^{-9}, respectively.[64,88] These three compounds are distributed relatively homogeneously in the stratosphere and are not confined to a narrow layer like hydrocarbons.[64,66]

Ethane, acetylene and diacetylene condense in the colder lower part of stratosphere[62] forming haze layers with an optical depth of about 0.01 in visible light.[89] Condensation occurs at approximately 14, 2.5 and 0.1 mbar for ethane, acetylene and diacetylene, respectively.[90,91] </ref> The concentration of hydrocarbons in the Uranian stratosphere is significantly lower than in the stratospheres of the other giant planets—the upper atmosphere of Uranus is very clean and transparent above the haze layers.[84] This depletion is caused by weak vertical mixing, and makes Uranus's stratosphere less opaque and, as a result, colder than those of other giant planets.[84,92] The hazes, like their parent hydrocarbons, are distributed unevenly across Uranus; at the solstice of 1986, when *Voyager 2* passed by the planet, they were concentrated near the sunlit pole, making it dark in ultraviolet light.[93]

Thermosphere and ionosphere

The outermost layer of the Uranian atmosphere, extending for thousands of kilometres, is the thermosphere/exosphere, which has a uniform temperature of around 800 to 850 K.[81,94] This is much higher than, for instance, the 420 K observed in the thermosphere of Saturn.[95] The heat sources necessary to sustain such high temperatures are not understood, since neither solar FUV/EUV radiation nor auroral activity can provide the necessary energy.[80,94] The weak cooling efficiency due to the depletion of hydrocarbons in the stratosphere may contribute to this phenomenon.[84] In addition to molecular hydrogen, the thermosphere contains a large proportion of free hydrogen atoms,[80] while helium is thought to be absent here, because it separates diffusively at lower altitudes.[96]

The thermosphere and upper part of the stratosphere contain a large concentration of ions and electrons, forming the ionosphere of Uranus.[97] Radio occultation observations by the *Voyager 2* spacecraft showed that the ionosphere lies between 1,000 and 10,000 km altitude and may include several narrow and dense layers between 1,000 and 3,500 km.[97,98] The electron density in the Uranian ionosphere is on average 10^4 cm^{-3},[99] reaching to as high as 10^5 cm^{-3} in the narrow layers in the stratosphere.[98] The ionosphere is mainly sustained by solar UV radiation and its density depends on the solar activity.[99,100] The auroral activity on Uranus is not as powerful as at Jupiter and Saturn and contributes little to the ionization.[101] </ref>[102] The high electron density may be in part caused by the low concentration of hydrocarbons in the stratosphere.[84]

One of the sources of information about the ionosphere and thermosphere comes from ground-based measurements of the intense mid-infrared (3–4 μm) emissions of the trihydrogen cation (H_3^+).[99,103] The total emitted power is $1–2 \times 10^{11}$ W—an order of magnitude higher than that the near-infrared hydrogen quadrupole emissions.[104] </ref>[105] Trihydrogen cation functions as one of main coolers of the ionosphere.[106]

The upper atmosphere of Uranus is the source of the far ultraviolet (90–140 nm) emissions known as *dayglow* or *electroglow*, which, like the H_3^+ IR radiation, emanates exclusively from the sunlit part of the planet. This phenomenon, which occurs in the thermospheres of all giant planets and was mysterious for a time after its discovery, is interpreted as a UV fluorescence of atomic and molecular hydrogen excited by solar radiation or by photoelectrons.[107]

Hydrogen corona

The upper part of the thermosphere, where the mean free path of the molecules exceeds the scale height,[108] T is temperature and $g_j \approx 8.9$ m/s^2 is the gravitational acceleration at the surface of Uranus. As the temperature varies from 53 K in the tropopause up to 800 K in the thermosphere, the scale height changes from 20 to 400 km. </ref> is called the exosphere.[109] The lower boundary of the Uranian exosphere, the exobase, is located at a height of about 6,500 km, or 1/4 of the planetary radius, above the surface.[109] The exosphere is unusually extended, reaching as far as several Uranian radii from the planet.[110,111] It is made mainly of hydrogen atoms and is often called the hydrogen corona of Uranus.[112] The high temperature and relatively high pressure at the base of the thermosphere explain in part why Uranus's exosphere is so vast.[113] </ref>[111] The number density of atomic hydrogen in the corona falls slowly with the distance from the planet, remaining as high a few hundred atoms per cm^3 at a few radii from Uranus.[114] The effects of this bloated exosphere include a drag on small particles orbiting Uranus, causing a general depletion of dust in the Uranian rings. The infalling dust in turn contaminates the upper atmosphere of the planet.[112]

Dynamics

Uranus has a relatively bland appearance, lacking broad colorful bands and large clouds prevalent on Jupiter and Saturn.[44,93] No discrete features were observed in Uranus's atmosphere before 1986.[40] The most conspicuous features on Uranus observed by *Voyager 2* were the dark low latitude region between –40° and –20° and bright southern polar cap.[93] The northern boundary of the cap was located at about –45° of latitude. The brightest zonal band was located near the edge of the cap at –50° to –45° and was then called a polar collar.[115] The southern polar cap, which existed at the time of the solstice in 1986, faded away in 1990s.[116] After the equinox in 2007, the southern polar collar started to fade away as well, while the northern polar collar located at 45° to 50° latitude (first appeared in 2007) have grown more conspicuous since then.[117]

The atmosphere of Uranus is calm compared to those of other giant planets. Only a limited number of small bright clouds at middle latitudes in both hemispheres[44] and one Uranus Dark Spot have been observed since 1986.[78] One of those bright cloud features, located at –34° of latitude and called *Berg*, probably existed continuously since at least 1986.[118] Nevertheless, the Uranian atmosphere has rather strong zonal winds blowing in the retrograde (counter to the rotation) direction near the equator, but switching to the prograde direction poleward of ±20° latitude.[119] The wind speeds are from –50 to –100 m/s at the

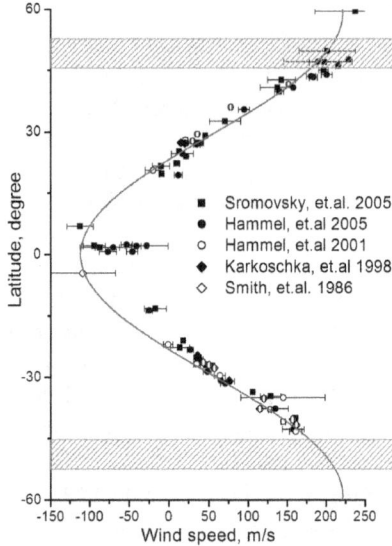

Figure 23: *Zonal wind speeds on Uranus. Shaded areas show the southern collar and its future northern counterpart. The red curve is a symmetrical fit to the data.*

equator increasing up to 240 m/s near 50° latitude.[116] The wind profile measured before the equinox of 2007 was slightly asymmetric with winds stronger in the southern hemisphere, although it turned out to be a seasonal effect as this hemisphere was continuously illuminated by the Sun before 2007.[116] After 2007 winds in the northern hemisphere accelerated while those in the southern one slowed down.

Uranus exhibits a considerable seasonal variation over its 84-year orbit. It is generally brighter near solstices and dimmer at equinoxes.[79] The variations are largely caused by changes in the viewing geometry: a bright polar region comes into view near solstices, while the dark equator is visible near equinoxes.[120] Still there exist some intrinsic variations of the reflectivity of the atmosphere: periodically fading and brightening polar caps as well as appearing and disappearing polar collars.[120]

References

<templatestyles src="Template:Refbegin/styles.css" />

- Adel, A.; Slipher, V. (1934). "The Constitution of the Atmospheres of the Giant Planets". *Physical Review*. **46** (10): 902. Bibcode: 1934PhRv... 46..902A[121]. doi: 10.1103/PhysRev.46.902[122].

- Atreya, Sushil K.; Wong, Ah-San (2005). "Coupled Clouds and Chemistry of the Giant Planets — A Case for Multiprobes"[123] (PDF). *Space Science Reviews*. **116**: 121–136. Bibcode: 2005SSRv..116..121A[124]. doi: 10.1007/s11214-005-1951-5[125].

- Bishop, J.; Atreya, S. K.; Herbert, F.; Romani, P. (December 1990). "Reanalysis of voyager 2 UVS occultations at Uranus: Hydrocarbon mixing ratios in the equatorial stratosphere"[126] (PDF). *Icarus*. **88** (2): 448–464. Bibcode: 1990Icar...88..448B[127]. doi: 10.1016/0019-1035(90)90094-P[128].

- Burgdorf, M.; Orton, G.; Vancleve, J.; Meadows, V.; Houck, J. (October 2006). "Detection of new hydrocarbons in Uranus' atmosphere by infrared spectroscopy". *Icarus*. **184** (2): 634–637. Bibcode: 2006Icar.. 184..634B[129]. doi: 10.1016/j.icarus.2006.06.006[130].

- Conrath, B.; Gautier, D.; Hanel, R.; Lindal, G.; Marten, A. (1987). "The Helium Abundance of Uranus from Voyager Measurements". *Journal of Geophysical Research*. **92** (A13): 15003–15010. Bibcode: 1987JGR.... 9215003C[131]. doi: 10.1029/JA092iA13p15003[132].

- Encrenaz, Thérèse (February 2003). "ISO observations of the giant planets and Titan: what have we learnt?". *Planetary and Space Science*. **51** (2): 89–103. Bibcode: 2003P&SS...51...89E[133]. doi: 10.1016/S0032-0633(02)00145-9[134].

- Encrenaz, T.; Drossart, P.; Orton, G.; Feuchtgruber, H.; Lellouch, E.; Atreya, S. K. (December 2003). "The rotational temperature and column density of H_3^+ in Uranus"[135] (PDF). *Planetary and Space Science*. **51** (14–15): 1013–1016. Bibcode: 2003P&SS...51.1013E[136]. doi: 10.1016/j.pss.2003.05.010[137].

- Encrenaz, T.; Lellouch, E.; Drossart, P.; Feuchtgruber, H.; Orton, G. S.; Atreya, S. K. (January 2004). "First detection of CO in Uranus"[138] (PDF). *Astronomy and Astrophysics*. **413** (2): L5–L9. Bibcode: 2004A&A... 413L...5E[139]. doi: 10.1051/0004-6361:20034637[140].

- Encrenaz, T. R. S. (January 2005). "Neutral Atmospheres of the Giant Planets: An Overview of Composition Measurements". *Space Science Reviews*. **116** (1–2): 99–119. Bibcode: 2005SSRv..116...99E[141]. doi: 10.1007/s11214-005-1950-6[142].

- Fegley, Bruce Jr.; Gautier, Daniel; Owen, Tobias; Prinn, Ronald G. (1991). "Spectroscopy and chemistry of the atmosphere of Uranus". In Bergstrahl, Jay T.; Miner, Ellis D.; Matthews, Mildred Shapley. *Uranus*[143] (PDF). University of Arizona Press. ISBN 978-0-8165-1208-9. OCLC 22625114[144].

- Feuchtgruber, H.; Lellouch, E.; Bézard, B.; Encrenaz, Th.; de Graauw, Th.; Davis, G. R. (January 1999). "Detection of HD in the atmospheres of Uranus and Neptune: a new determination of the D/H ratio". *Astron-*

omy and Astrophysics. **341**: L17–L21. Bibcode: 1999A&A...341L.. 17F[145].

- Fry, Patrick M.; Sromovsky, L. A. (September 2009). *Implications of New Methane Absorption Coefficients on Uranus Vertical Structure Derived from Near-IR Spectra*. DPS meeting #41, #14.06. American Astronomical Society. Bibcode: 2009DPS....41.1406F[146].
- Hammel, H. B.; Lockwood, G. W. (January 2007). "Long-term atmospheric variability on Uranus and Neptune". *Icarus*. **186** (1): 291–301. Bibcode: 2007Icar..186..291H[147]. doi: 10.1016/j.icarus.2006.08.027[148].
- Hammel, H. B.; Sromovsky, L. A.; Fry, P. M.; Rages, K.; Showalter, M.; de Pater, I.; van Dam, M. A.; LeBeau, R. P.; Deng, X. (May 2009). "The Dark Spot in the atmosphere of Uranus in 2006: Discovery, description, and dynamical simulations"[149] (PDF). *Icarus*. **201** (1): 257–271. Bibcode: 2009Icar..201..257H[150]. doi: 10.1016/j.icarus.2008.08.019[151]. Archived from the original[152] (PDF) on 2011-07-19.
- Hanel, R.; Conrath, B.; Flasar, F. M.; Kunde, V.; Maguire, W.; Pearl, J.; Pirraglia, J.; Samuelson, R.; Cruikshank, D. (4 July 1986). "Infrared Observations of the Uranian System". *Science*. **233** (4759): 70–74. Bibcode: 1986Sci...233...70H[153]. doi: 10.1126/science.233.4759.70[154]. PMID 17812891[155].
- Herbert, F.; Sandel, B. R.; Yelle, R. V.; Holberg, J. B.; Broadfoot, A. L.; Shemansky, D. E.; Atreya, S. K.; Romani, P. N. (December 30, 1987). "The Upper Atmosphere of Uranus: EUV Occultations Observed by Voyager 2"[156] (PDF). *Journal of Geophysical Research*. **92** (A13): 15,093–15,109. Bibcode: 1987JGR....9215093H[157]. doi: 10.1029/JA092iA13p15093[158].
- Herbert, F.; Hall, D. T. (May 1996). "Atomic hydrogen corona of Uranus". *Journal of Geophysical Research*. **101** (A5): 10,877–10,885. Bibcode: 1996JGR...10110877H[159]. doi: 10.1029/96JA00427[160].
- Herbert, Floyd; Sandel, Bill R. (August–September 1999). "Ultraviolet observations of Uranus and Neptune". *Planetary and Space Science*. **47** (8–9): 1,119–1,139. Bibcode: 1999P&SS...47.1119H[161]. doi: 10.1016/S0032-0633(98)00142-1[162].
- Herzberg, G. (May 1952). "Spectroscopic evidence of molecular hydrogen in the atmospheres of Uranus and Neptune". *The Astrophysical Journal*. **115**: 337–340. Bibcode: 1952ApJ...115..337H[163]. doi: 10.1086/145552[164].
- Huggins, William (June 1889). "The spectrum of Uranus". *Monthly Notices of the Royal Astronomical Society*. **49**: 404. Bibcode: 1889MN-RAS..49Q.404H[165]. doi: 10.1093/mnras/49.8.403a[166].
- Irwin, P. G. J.; Teanby, N. A.; Davis, G. R. (2007-08-10). "Latitudinal Variations in Uranus' Vertical Cloud Structure from UKIRT UIST

Observations". *The Astrophysical Journal.* The American Astronomical Society. **665** (1): L71–L74. Bibcode: 2007ApJ...665L..71I[167]. doi: 10.1086/521189[168].

• Irwin, P. G. J.; Teanby, N. A.; Davis, G. R. (August 2010). "Revised vertical cloud structure of Uranus from UKIRT/UIST observations and changes seen during Uranus' Northern Spring Equinox from 2006 to 2008: Application of new methane absorption data and comparison with Neptune". *Icarus.* **208** (2): 913–926. Bibcode: 2010Icar..208..913I[169]. doi: 10.1016/j.icarus.2010.03.017[170].

• Kuiper, G. P. (May 1949). "New absorptions in the Uranian atmosphere". *The Astrophysical Journal.* **109**: 540–541. Bibcode: 1949ApJ... 109..540K[171]. doi: 10.1086/145161[172].

• Lam, H. A.; Miller, S.; Joseph, R. D.; Geballe, T. R.; Trafton, L. M.; Tennyson, J.; Ballester, G. E. (1997-01-01). "Variation in the H_3^+ Emission of Uranus"[173] (PDF). *The Astrophysical Journal.* The American Astronomical Society. **474** (1): L73–L76. Bibcode: 1997ApJ...474L..73L[174]. doi: 10.1086/310424[175].

• Lindal, G. F.; Lyons, J. R.; Sweetnam, D. N.; Eshleman, V. R.; Hinson, D. P.; Tyler, G. L. (December 30, 1987). "The Atmosphere of Uranus: Results of Radio Occultation Measurements with Voyager 2". *Journal of Geophysical Research.* American Geophysical Union. **92** (A13): 14,987–15,001. Bibcode: 1987JGR....9214987L[176]. doi: 10.1029/JA092iA13p14987[177].

• Lockyer, J. N. (June 1889). "Note on the Spectrum of Uranus". *Astronomische Nachrichten.* **121**: 369. Bibcode: 1889AN....121..369L[178]. doi: 10.1002/asna.18891212402[179].

• Lodders, Katharina (July 10, 2003). "Solar System Abundances and Condensation Temperatures of the Elements"[180] (PDF). *The Astrophysical Journal.* The American Astronomical Society. **591** (2): 1220–1247. Bibcode: 2003ApJ...591.1220L[181]. doi: 10.1086/375492[182].

• Lunine, Jonathan I. (September 1993). "The Atmospheres of Uranus and Neptune". *Annual Review of Astronomy and Astrophysics.* **31**: 217–263. Bibcode: 1993ARA&A..31..217L[183]. doi: 10.1146/annurev.aa. 31.090193.001245[184].

• Miller, Steven; Achilleos, Nick; Ballester, Gilda E.; Geballe, Thomas R.; Joseph, Robert D.; Prangé, Renee; Rego, Daniel; Stallard, Tom; Tennyson, Jonathan; Trafton, Laurence M.; Waite, J. Hunter Jr (15 September 2000). "The role of H_3^+ in planetary atmospheres"[185] (PDF). *Philosophical Transactions of the Royal Society A: Mathematical, Physical and Engineering Sciences.* **358** (1774): 2485–2502. doi: 10.1098/rsta. 2000.0662[186].

• Miller, Steve; Aylward, Alan; Millward, George (January 2005). "Giant

Planet Ionospheres and Thermospheres: The Importance of Ion-Neutral Coupling". *Space Science Reviews*. **116** (1–2): 319–343. Bibcode: 2005SSRv..116..319M[187]. doi: 10.1007/s11214-005-1960-4[188].

- Rages, K. A.; Hammel, H. B.; Friedson, A. J. (11 September 2004). "Evidence for temporal change at Uranus' south pole". *Icarus*. **172** (2): 548–554. Bibcode: 2004Icar..172..548R[189]. doi: 10.1016/j.icarus. 2004.07.009[190].

- de Pater, I.; Romani, P. N.; Atreya, S. K. (December 1989). "Uranius Deep Atmosphere Revealed"[191] (PDF). *Icarus*. **82** (2): 288–313. Bibcode: 1989Icar...82..288D[192]. doi: 10.1016/0019-1035(89)90040-7[193].

- de Pater, Imke; Romani, Paul N.; Atreya, Sushil K. (June 1991). "Possible microwave absorption by H_2S gas in Uranus' and Neptune's atmospheres"[194] (PDF). *Icarus*. **91** (2): 220–233. Bibcode: 1991Icar... 91..220D[195]. doi: 10.1016/0019-1035(91)90020-T[196].

- Pearl, J. C.; Conrath, B. J.; Hanel, R. A.; Pirraglia, J. A.; Coustenis, A. (March 1990). "The albedo, effective temperature, and energy balance of Uranus, as determined from Voyager IRIS data". *Icarus*. **84** (1): 12–28. Bibcode: 1990Icar...84...12P[197]. doi: 10.1016/0019-1035(90)90155-3[198].

- Pollack, James B.; Rages, Kathy; Pope, Shelly K.; Tomasko, Martin G.; Romani, Paul N.; Atreya, Sushil K. (December 30, 1987). "Nature of the Stratospheric Haze on Uranus: Evidence for Condensed Hydrocarbons"[199] (PDF). *Journal of Geophysical Research*. **92** (A13): 15,037–15,065. Bibcode: 1987JGR....9215037P[200]. doi: 10.1029/ JA092iA13p15037[201].

- Smith, B. A. (October 1984). "Near infrared imaging of Uranus and Neptune". *In JPL Uranus and Neptune*. **2330**: 213–223. Bibcode: 1984NASCP2330..213S[202].

- Smith, B. A.; Soderblom, L. A.; Beebe, A.; Bliss, D.; Boyce, J. M.; Brahic, A.; Briggs, G. A.; Brown, R. H.; Collins, S. A. (4 July 1986). "Voyager 2 in the Uranian System: Imaging Science Results". *Science*. **233** (4759): 43–64. Bibcode: 1986Sci...233...43S[203]. doi: 10.1126/ science.233.4759.43[204]. PMID 17812889[205].

- Sromovsky, L. A.; Fry, P. M. (December 2005). "Dynamics of cloud features on Uranus". *Icarus*. **179** (2): 459–484. arXiv: 1503.03714[206] ⓐ. Bibcode: 2005Icar..179..459S[207]. doi: 10.1016/j.icarus.2005.07.022[208].

- Sromovsky, L. A.; Irwin, P. G. J.; Fry, P. M. (June 2006). "Near-IR methane absorption in outer planet atmospheres: Improved models of temperature dependence and implications for Uranus cloud structure". *Icarus*. **182** (2): 577–593. Bibcode: 2006Icar..182..577S[209]. doi: 10.1016/j.icarus.2006.01.008[210].

- Sromovsky, L. A.; Fry, P. M.; Hammel, H. B.; Ahue, W. M.; de Pater, I.; Rages, K. A.; Showalter, M. R.; van Dam, M. A. (September 2009).

"Uranus at equinox: Cloud morphology and dynamics". *Icarus*. **203** (1): 265–286. arXiv: 1503.01957[211] ə. Bibcode: 2009Icar..203..265S[212]. doi: 10.1016/j.icarus.2009.04.015[213].

- Summers, M. E.; Strobel, D. F. (November 1, 1989). "Photochemistry of the atmosphere of Uranus". *The Astrophysical Journal*. **346**: 495–508. Bibcode: 1989ApJ...346..495S[214]. doi: 10.1086/168031[215].

- Stone, E. C. (December 30, 1987). "The Voyager 2 Encounter with Uranus". *Journal of Geophysical Research*. **92** (A13): 14,873–14,876. Bibcode: 1987JGR....9214873S[216]. doi: 10.1029/JA092iA13p14873[217].

- Trafton, L. M.; Miller, S.; Geballe, T. R.; Tennyson, J.; Ballester, G. E. (October 1999). "H_2 Quadrupole and $H_3{}^+$ Emission from Uranus: The Uranian Thermosphere, Ionosphere, and Aurora". *The Astrophysical Journal*. **524** (2): 1,059–1,083. Bibcode: 1999ApJ...524.1059T[218]. doi: 10.1086/307838[219].

- Tyler, G. L.; Sweetnam, D. N.; Anderson, J. D.; Campbell, J. K.; Eshleman, V. R.; Hinson, D. P.; Levy, G. S.; Lindal, G. F.; Marouf, E. A.; Simpson, R. A. (4 July 1986). "Voyager 2 Radio Science Observations of the Uranian System: Atmosphere, Rings, and Satellites". *Science*. **233** (4759): 79–84. Bibcode: 1986Sci...233...79T[220]. doi: 10.1126/science.233.4759.79[221]. PMID 17812893[222].

- Young, L. (2001). "Uranus after Solstice: Results from the 1998 November 6 Occultation"[223] (PDF). *Icarus*. **153** (2): 236–247. Bibcode: 2001Icar..153..236Y[224]. doi: 10.1006/icar.2001.6698[225].

External links

ə Media related to Uranus (atmosphere) at Wikimedia Commons

Climate

Climate of Uranus

The **climate of Uranus** is heavily influenced by both its lack of internal heat, which limits atmospheric activity, and by its extreme axial tilt, which induces intense seasonal variation. Uranus' atmosphere is remarkably bland in comparison to the other gas giants which it otherwise closely resembles.[226] When *Voyager 2* flew by Uranus in 1986, it observed a total of ten cloud features across the entire planet.[227,228] Later observations from the ground or by the Hubble Space Telescope made in the 1990s and the 2000s revealed bright clouds in the northern (winter) hemisphere. In 2006 a dark spot similar to the Great Dark Spot on Neptune was detected.[229]

Banded structure, winds and clouds

In 1986 Voyager 2 discovered that the visible southern hemisphere of Uranus can be subdivided into two regions: a bright polar cap and dark equatorial bands (see figure on the right).[230] Their boundary is located at about −45 degrees of latitude. A narrow band straddling the latitudinal range from −45 to −50 degrees is the brightest large feature on Uranus's visible surface.[230,231] It is called a southern "collar". The cap and collar are thought to be a dense region of methane clouds located within the pressure range of 1.3 to 2 bar.[232] Unfortunately Voyager 2 arrived during the height of Uranus's southern summer and could not observe the northern hemisphere. However, at the end of 1990s and the beginning of the twenty-first century, when the northern polar region came into view, Hubble Space Telescope (HST) and Keck telescope initially observed neither a collar nor a polar cap in the northern hemisphere.[231] So Uranus appeared to be asymmetric: bright near the south pole and uniformly dark in the region north of the southern collar.[231] In 2007, however, when Uranus passed its equinox, the southern collar almost disappeared, whereas a

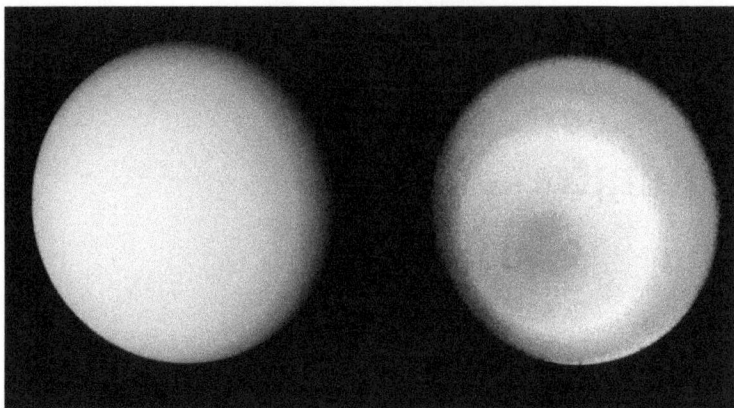

Figure 24: *Uranus' southern hemisphere in approximate natural colour (left) and in higher wavelengths (right), showing its faint cloud bands and atmospheric "hood" as seen by Voyager 2*

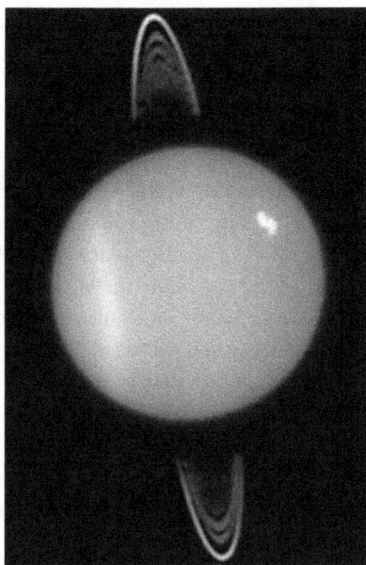

Figure 25: *Uranus in 2005. Rings, southern collar and a light cloud in the northern hemisphere are visible.*

faint northern collar emerged near 45 degrees of latitude.[233] The visible latitudinal structure of Uranus is different from that of Jupiter and Saturn, which demonstrate multiple narrow and colorful bands.[226]

In addition to large-scale banded structure, Voyager 2 observed ten small bright clouds, most lying several degrees to the north from the collar.[230] In all other respects Uranus looked like a dynamically dead planet in 1986. However, in the 1990s the number of the observed bright cloud features grew considerably.[226] The majority of them was found in the northern hemisphere as it started to become visible.[226] The common though incorrect explanation of this fact was that bright clouds are easier to identify in its dark part, whereas in the southern hemisphere the bright collar masks them.[234] Nevertheless, there are differences between the clouds of each hemisphere. The northern clouds are smaller, sharper and brighter.[235] They appear to lie at a higher altitude, which is connected to fact that until 2004 (see below) no southern polar cloud had been observed at the wavelength 2.2 micrometres,[235] which is sensitive to the methane absorption, whereas northern clouds have been regularly observed in this wavelength band. The lifetime of clouds spans several orders of magnitude. Some small clouds live for hours, whereas at least one southern cloud has persisted since the Voyager flyby.[226,228] Recent observation also discovered that cloud-features on Uranus have a lot in common with those on Neptune, although the weather on Uranus is much calmer.[226]

Uranus Dark Spot

The dark spots common on Neptune had never been observed on Uranus before 2006, when the first such feature was imaged.[236] In that year observations from both Hubble Space Telescope and Keck Telescope revealed a small dark spot in the northern (winter) hemisphere of Uranus. It was located at the latitude of about $28 \pm 1°$ and measured approximately $2°$ (1300 km) in latitude and $5°$ (2700 km) in longitude.[229] The feature called Uranus Dark Spot (UDS) moved in the prograde direction relative Uranus's rotation with an average speed of 43.1 ± 0.1 m/s, which is almost 20 m/s faster than the speed of clouds at the same latitude.[229] The latitude of UDS was approximately constant. The feature was variable in size and appearance and was often accompanied by a bright white clouds called Bright Companion (BC), which moved with nearly the same speed as UDS itself.[229]

The behavior and appearance of UDS and its bright companion were similar to Neptunian Great Dark Spots (GDS) and their bright companions, respectively, though UDS was significantly smaller. This similarity suggests that they have the same origin. GDS were hypothesized to be anticyclonic vortices in the atmosphere of Neptune, whereas their bright companions were thought to be methane clouds formed in places, where the air is rising (orographic clouds).[229]

Figure 26: *The first dark spot observed on Uranus.*
Image was obtained by ACS on HST in 2006.

UDS is supposed to have a similar nature, although it looked differently from GDS at some wavelengths. Although GDS had the highest contrast at 0.47 μm, UDS was not visible at this wavelength. On the other hand, UDS demonstrated the highest contrast at 1.6 μm, where GDS were not detected.[229] This implies that dark spots on the two ice giants are located at somewhat different pressure levels—the Uranian feature probably lies near 4 bar. The dark color of UDS (as well as GDS) may be caused by thinning of the underlying hydrogen sulfide or ammonium hydrosulfide clouds.[229]

The emergence of a dark spot on the hemisphere of Uranus that was in darkness for many years indicates that near equinox Uranus entered a period of elevated weather activity.[229]

Winds

The tracking of numerous cloud features allowed determination of zonal winds blowing in the upper troposphere of Uranus.[226] At the equator winds are retrograde, which means that they blow in the reverse direction to the planetary rotation. Their speeds are from –100 to –50 m/s.[226,231] Wind speeds increase with the distance from the equator, reaching zero values near ±20° latitude, where the troposphere's temperature minimum is located.[226,237] Closer to the

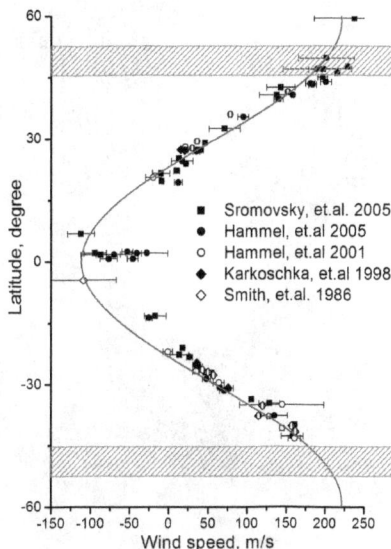

Figure 27: *Zonal wind speeds on Uranus. Shaded areas show the southern collar and its future northern counterpart. The red curve is a symmetrical fit to the data.*

poles, the winds shift to a prograde direction, flowing with its rotation. Wind speeds continue to increase reaching maxima at ±60° latitude before falling to zero at the poles.[226] Wind speeds at –40° latitude range from 150 to 200 m/ s. Because the collar obscures all clouds below that parallel, speeds between it and the south pole are impossible to measure.[226] In contrast, in the northern hemisphere maximum speeds as high as 240 m/s are observed near +50 degrees of latitude.[226,231] These speeds sometimes lead to incorrect assertions that winds are faster in the northern hemisphere. In fact, latitude per latitude, winds are slightly slower in the northern part of Uranus, especially at the mid-latitudes from ±20 to ±40 degrees.[226] There is currently no agreement about whether any changes in wind speed have occurred since 1986,[226,231,238] and nothing is known about much slower meridional winds.[226]

Seasonal variation

Determining the nature of this seasonal variation is difficult because good data on Uranus's atmosphere has existed for less than 84 Earth years, or one full Uranian year. A number of discoveries have however been made. Photometry over the course of half a Uranian year (beginning in the 1950s) has shown regular variation in the brightness in two spectral bands, with maxima occurring

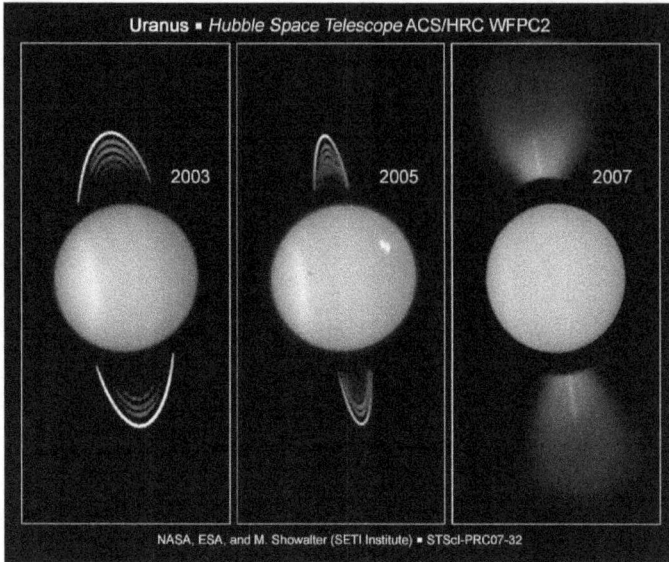

Figure 28: *HST images show changes in the atmosphere
of Uranus as it approaches its equinox (right image)*

at the solstices and minima occurring at the equinoxes.[239] A similar periodic
variation, with maxima at the solstices, has been noted in microwave measurements of the deep troposphere begun in the 1960s.[240] Stratospheric temperature measurements beginning in the 1970s also showed maximum values near
1986 solstice.[241]

The majority of this variability is believed to occur due to changes in the viewing geometry. Uranus is an oblate spheroid, which causes its visible area to
become larger when viewed from the poles. This explains in part its brighter
appearance at solstices.[239] Uranus is also known to exhibit strong meridional
variations in albedo (see above).[234] For instance, the south polar region of
Uranus is much brighter than the equatorial bands.[230] In addition, both poles
demonstrate elevated brightness in the microwave part of the spectrum,[242]
whereas the polar stratosphere is known to be cooler than the equatorial one.[241]
So seasonal change seems to happen as follows: poles, which are bright both in
visible and microwave spectral bands, come into the view at solstices resulting
in brighter planet, whereas the dark equator is visible mainly near equinoxes
resulting in darker planet.[234] In addition, occultations at solstices probe hotter
equatorial stratosphere.[241]

However, there are some reasons to believe that seasonal changes are happening in Uranus. Although Uranus is known to have a bright south polar

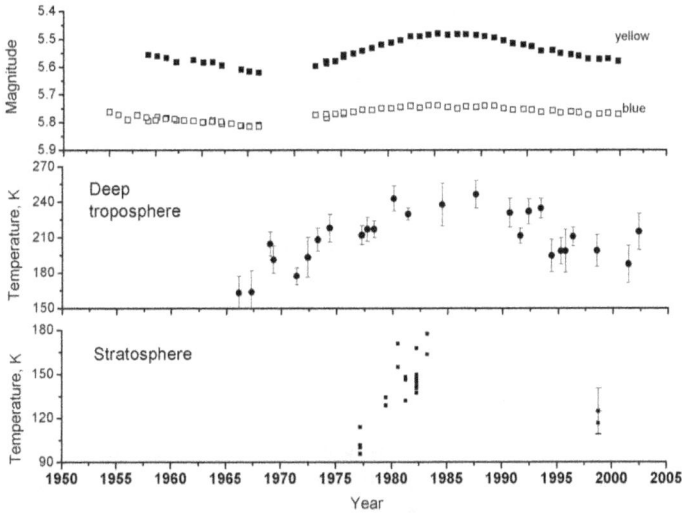

Figure 29: *The visible magnitude of Uranus in two spectral bands (upper graph)*[239] *adjusted for the distance, effective microwave temperature (middle graph) and the stratospheric temperature (lower graph).*[240] *Blue band is centered at 470 nm, yellow at 550 nm.*

region, the north pole is fairly dim, which is incompatible with the model of the seasonal change outlined above.[243] During its previous northern solstice in 1944, Uranus displayed elevated levels of brightness, which suggests that the north pole was not always so dim.[239] This information implies that the visible pole brightens some time before the solstice and darkens after the equinox.[243] Detailed analysis of the visible and microwave data revealed that the periodical changes of brightness are not completely symmetrical around the solstices, which also indicates a change in the albedo patterns.[243] In addition, the microwave data showed increases in pole–equator contrast after the 1986 solstice.[242] Finally in the 1990s, as Uranus moved away from its solstice, Hubble and ground-based telescopes revealed that the south polar cap darkened noticeably (except the southern collar, which remained bright),[232] whereas the northern hemisphere demonstrated increasing activity,[228] such as cloud formations and stronger winds, having bolstered expectations that it would brighten soon.[235] In particular, an analog of the bright polar collar present in its southern hemisphere at −45° was expected to appear in its northern part.[243] This indeed happened in 2007 when Uranus passed an equinox: a faint northern polar collar arose, whereas the southern collar became nearly invisible, although the zonal wind profile remained asymmetric, with northern winds being slightly slower than southern.[233]

The mechanism of physical changes is still not clear.[243] Near the summer and winter solstices, Uranus's hemispheres lie alternately either in full glare of the Sun's rays or facing deep space. The brightening of the sunlit hemisphere is thought to result from the local thickening of the methane clouds and haze layers located in the troposphere.[232] The bright collar at −45° latitude is also connected with methane clouds.[232] Other changes in the southern polar region can be explained by changes in the lower cloud layers.[232] The variation of the microwave emission from Uranus is probably caused by changes in the deep tropospheric circulation, because thick polar clouds and haze may inhibit convection.[242]

For a short period in Autumn 2004, a number of large clouds appeared in the Uranian atmosphere, giving it a Neptune-like appearance.[235,244] Observations included record-breaking wind speeds of 824 km/h and a persistent thunderstorm referred to as "Fourth of July fireworks".[228] Why this sudden upsurge in activity should be occurring is not fully known, but it appears that Uranus's extreme axial tilt results in extreme seasonal variations in its weather.[236,243]

Circulation models

Several solutions have been proposed to explain the calm weather on Uranus. One proposed explanation for this dearth of cloud features is that Uranus's internal heat appears markedly lower than that of the other giant planets; in astronomical terms, it has a low internal thermal flux.[226,237] Why Uranus's heat flux is so low is still not understood. Neptune, which is Uranus's near twin in size and composition, radiates 2.61 times as much energy into space as it receives from the Sun.[226] Uranus, by contrast, radiates hardly any excess heat at all. The total power radiated by Uranus in the far infrared (i.e. heat) part of the spectrum is 1.06 ± 0.08 times the solar energy absorbed in its atmosphere.[245,246] In fact, Uranus's heat flux is only 0.042 ± 0.047 W/m², which is lower than the internal heat flux of Earth of about 0.075 W/m².[245] The lowest temperature recorded in Uranus's tropopause is 49 K (−224 °C), making Uranus the coldest planet in the Solar System, colder than Neptune.[245,246]

Another hypothesis states that when Uranus was "knocked over" by the supermassive impactor which caused its extreme axial tilt, the event also caused it to expel most of its primordial heat, leaving it with a depleted core temperature. Another hypothesis is that some form of barrier exists in Uranus's upper layers which prevents the core's heat from reaching the surface.[247] For example, convection may take place in a set of compositionally different layers, which may inhibit the upward heat transport.[245,246]

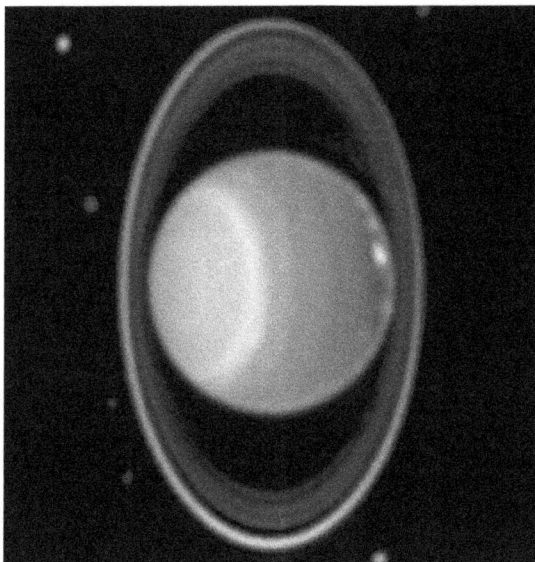

Figure 30: *HST image of Uranus taken in 1998 showing clouds in the northern hemisphere*

References

Sources

- Devitt, Terry (10 November 2004). "Keck zooms in on the weird weather of Uranus"[248]. University of Wisconsin–Madison. Retrieved 2012-03-10.
- Hammel, H. B.; De Pater, I.; Gibbard, S. G.; Lockwood, G. W.; Rages, K. (June 2005). "Uranus in 2003: Zonal winds, banded structure, and discrete features"[249] (PDF). *Icarus*. **175** (2): 534–545. Bibcode: 2005Icar.. 175..534H[250]. doi: 10.1016/j.icarus.2004.11.012[251].
- Hammel, H. B.; Depater, I.; Gibbard, S. G.; Lockwood, G. W.; Rages, K. (May 2005). "New cloud activity on Uranus in 2004: First detection of a southern feature at 2.2 μm"[252] (PDF). *Icarus*. **175** (1): 284–288. Bibcode: 2005Icar..175..284H[253]. doi: 10.1016/j.icarus.2004.11.016[254].
- Hammel, H. B.; Lockwood, G. W. (January 2007). "Long-term atmospheric variability on Uranus and Neptune". *Icarus*. **186** (1): 291–301. Bibcode: 2007Icar..186..291H[255]. doi: 10.1016/j.icarus.2006.08.027[256].
- Hammel, H. B.; Rages, K.; Lockwood, G. W.; Karkoschka, E.; de Pater, I. (October 2001). "New Measurements of the Winds of Uranus". *Icarus*. **153** (2): 229–235. Bibcode: 2001Icar..153..229H[257]. doi: 10.1006/icar. 2001.6689[258].

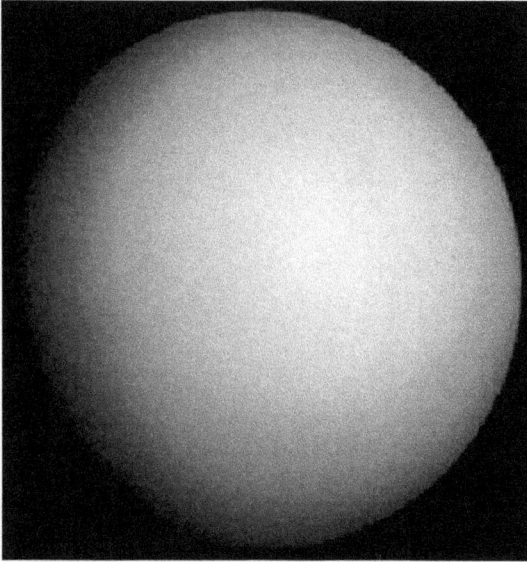

Figure 31: *The greenish color of Uranus's atmosphere is due to methane and high-altitude photochemical smog. Voyager 2 acquired this view of the seventh planet while departing the Uranian system in late January 1986. This image looks at Uranus approximately along its rotational pole.*

- Hammel, H. B.; Sromovsky, L. A.; Fry, P. M.; Rages, K.; Showalter, M.; de Pater, I.; van Dam, M. A.; LeBeau, R. P.; Deng, X. (May 2009). "The Dark Spot in the atmosphere of Uranus in 2006: Discovery, description, and dynamical simulations"[259] (PDF). *Icarus*. **201** (1): 257–271. Bibcode: 2009Icar..201..257H[260]. doi: 10.1016/j.icarus.2008.08.019[261]. Archived from the original[262] (PDF) on 2011-07-19.
- Hanel, R.; Conrath, B.; Flasar, F. M.; Kunde, V.; Maguire, W.; Pearl, J.; Pirraglia, J.; Samuelson, R.; Cruikshank, D. (4 July 1986). "Infrared Observations of the Uranian System". *Science*. **233** (4759): 70–74. Bibcode: 1986Sci...233...70H[263]. doi: 10.1126/science.233.4759.70[264]. PMID 17812891[265].
- Hofstadter, M. D.; Butler, B. J. (September 2003). "Seasonal change in the deep atmosphere of Uranus". *Icarus*. **165** (1): 168–180. Bibcode: 2003Icar..165..168H[266]. doi: 10.1016/S0019-1035(03)00174-X[267].
- Karkoschka, Erich (May 2001). "Uranus' Apparent Seasonal Variability in 25 HST Filters". *Icarus*. **151** (1): 84–92. Bibcode: 2001Icar.. 151...84K[268]. doi: 10.1006/icarus.2001.6599[269].

- Klein, M. J.; Hofstadter, M. D. (September 2006). "Long-term variations in the microwave brightness temperature of the Uranus atmosphere". *Icarus.* **184** (1): 170–180. Bibcode: 2006Icar..184..170K[270]. doi: 10.1016/j.icarus.2006.04.012[271].
- Lakdawalla, Emily (11 November 2004). "No Longer Boring: 'Fireworks' and Other Surprises at Uranus Spotted Through Adaptive Optics"[272]. *Planetary News: Observing from Earth.* The Planetary Society. Retrieved 2012-03-10.
- Lockwood, G. W.; Jerzykiewicz, M. A. A. (February 2006). "Photometric variability of Uranus and Neptune, 1950–2004". *Icarus.* **180** (2): 442–452. Bibcode: 2006Icar..180..442L[273]. doi: 10.1016/j.icarus. 2005.09.009[274].
- Lunine, Jonathan I. (September 1993). "The Atmospheres of Uranus and Neptune". *Annual Review of Astronomy and Astrophysics.* **31**: 217–263. Bibcode: 1993ARA&A..31..217L[275]. doi: 10.1146/annurev.aa. 31.090193.001245[276].
- Pearl, J. C.; Conrath, B. J.; Hanel, R. A.; Pirraglia, J. A.; Coustenis, A. (March 1990). "The albedo, effective temperature, and energy balance of Uranus, as determined from Voyager IRIS data". *Icarus.* **84** (1): 12–28. Bibcode: 1990Icar...84...12P[277]. doi: 10.1016/0019-1035(90)90155-3[278]. ISSN 0019-1035[279].
- Podolak, M.; Weizman, A.; Marley, M. (December 1995). "Comparative models of Uranus and Neptune". *Planetary and Space Science.* **43** (12): 1517–1522. Bibcode: 1995P&SS...43.1517P[280]. doi: 10.1016/0032-0633(95)00061-5[281].
- Rages, K. A.; Hammel, H. B.; Friedson, A. J. (11 September 2004). "Evidence for temporal change at Uranus' south pole". *Icarus.* **172** (2): 548–554. Bibcode: 2004Icar..172..548R[282]. doi: 10.1016/j.icarus. 2004.07.009[283].
- Smith, B. A.; Soderblom, L. A.; Beebe, A.; Bliss, D.; Boyce, J. M.; Brahic, A.; Briggs, G. A.; Brown, R. H.; Collins, S. A. (4 July 1986). "Voyager 2 in the Uranian System: Imaging Science Results". *Science.* **233** (4759): 43–64. Bibcode: 1986Sci...233...43S[284]. doi: 10.1126/ science.233.4759.43[285]. PMID 17812889[286].
- Sromovsky, L. A.; Fry, P. M. (December 2005). "Dynamics of cloud features on Uranus". *Icarus.* **179** (2): 459–484. arXiv: 1503.03714[287] ⊜. Bibcode: 2005Icar..179..459S[288]. doi: 10.1016/j.icarus.2005.07.022[289].
- Sromovsky, L. A.; Fry, P. M.; Hammel, H. B.; Ahue, W. M.; de Pater, I.; Rages, K. A.; Showalter, M. R.; van Dam, M. A. (September 2009). "Uranus at equinox: Cloud morphology and dynamics". *Icarus.* **203** (1): 265–286. arXiv: 1503.01957[290] ⊜. Bibcode: 2009Icar..203..265S[291]. doi: 10.1016/j.icarus.2009.04.015[292].

- Sromovsky, L.; Fry, P.; Hammel, H.; Rages, K. (September 28, 2006). "Hubble Discovers a Dark Cloud in the Atmosphere of Uranus"[293] (PDF). PHYSorg.com. Retrieved 2012-02-27.
- Young, L. (2001). "Uranus after Solstice: Results from the 1998 November 6 Occultation"[294] (PDF). *Icarus*. **153** (2): 236–247. Bibcode: 2001Icar..153..236Y[295]. doi: 10.1006/icar.2001.6698[296].

External links

- What is the Temperature of Uranus?[297] by Nola Taylor
- Uranus Facts[298]

Formation

Formation and evolution of the Solar System

The formation and evolution of the Solar System began 4.6 billion years ago with the gravitational collapse of a small part of a giant molecular cloud. Most of the collapsing mass collected in the center, forming the Sun, while the rest flattened into a protoplanetary disk out of which the planets, moons, asteroids, and other small Solar System bodies formed.

This model, known as the nebular hypothesis and now refined as the Nice model (2005), was first developed in the 18th century by Emanuel Swedenborg, Immanuel Kant, and Pierre-Simon Laplace. Its subsequent development has interwoven a variety of scientific disciplines including astronomy, physics, geology, and planetary science. Since the dawn of the space age in the 1950s and the discovery of extrasolar planets in the 1990's, the model has been both challenged and refined to account for new observations.

The Solar System has evolved considerably since its initial formation. Many moons have formed from circling discs of gas and dust around their parent planets, while other moons are thought to have formed independently and later been captured by their planets. Still others, such as Earth's Moon, may be the result of giant collisions. Collisions between bodies have occurred continually up to the present day and have been central to the evolution of the Solar System. The positions of the planets often shifted due to gravitational interactions. This planetary migration is now thought to have been responsible for much of the Solar System's early evolution.

In roughly 5 billion years, the Sun will cool and expand outward to many times its current diameter (becoming a red giant), before casting off its outer layers as a planetary nebula and leaving behind a stellar remnant known as a white dwarf. In the far distant future, the gravity of passing stars will gradually reduce the

Figure 32: *Artist's conception of a protoplanetary disk*

Sun's retinue of planets. Some planets will be destroyed, others ejected into interstellar space. Ultimately, over the course of tens of billions of years, it is likely that the Sun will be left with none of the original bodies in orbit around it.

History

Ideas concerning the origin and fate of the world date from the earliest known writings; however, for almost all of that time, there was no attempt to link such theories to the existence of a "Solar System", simply because it was not generally thought that the Solar System, in the sense we now understand it, existed. The first step toward a theory of Solar System formation and evolution was the general acceptance of heliocentrism, which placed the Sun at the centre of the system and the Earth in orbit around it. This concept had developed for millennia (Aristarchus of Samos had suggested it as early as 250 BC), but was not widely accepted until the end of the 17th century. The first recorded use of the term "Solar System" dates from 1704.

The current standard theory for Solar System formation, the nebular hypothesis, has fallen into and out of favour since its formulation by Emanuel Swedenborg, Immanuel Kant, and Pierre-Simon Laplace in the 18th century. The most significant criticism of the hypothesis was its apparent inability to explain the Sun's relative lack of angular momentum when compared to the planets. However, since the early 1980s studies of young stars have shown them to be surrounded by cool discs of dust and gas, exactly as the nebular hypothesis predicts, which has led to its re-acceptance.

Figure 33: *Pierre-Simon Laplace, one of the originators of the nebular hypothesis*

Understanding of how the Sun is expected to continue to evolve required an understanding of the source of its power. Arthur Stanley Eddington's confirmation of Albert Einstein's theory of relativity led to his realisation that the Sun's energy comes from nuclear fusion reactions in its core, fusing hydrogen into helium. In 1935, Eddington went further and suggested that other elements also might form within stars. Fred Hoyle elaborated on this premise by arguing that evolved stars called red giants created many elements heavier than hydrogen and helium in their cores. When a red giant finally casts off its outer layers, these elements would then be recycled to form other star systems.

Formation

Pre-solar nebula

The nebular hypothesis says that the Solar System formed from the gravitational collapse of a fragment of a giant molecular cloud. The cloud was about 20 parsec (65 light years) across, while the fragments were roughly 1 parsec (three and a quarter light-years) across. The further collapse of the fragments led to the formation of dense cores 0.01–0.1 pc (2,000–20,000 AU) in size.[299] One of these collapsing fragments (known as the *pre-solar nebula*) formed what became the Solar System. The composition of this region with a mass

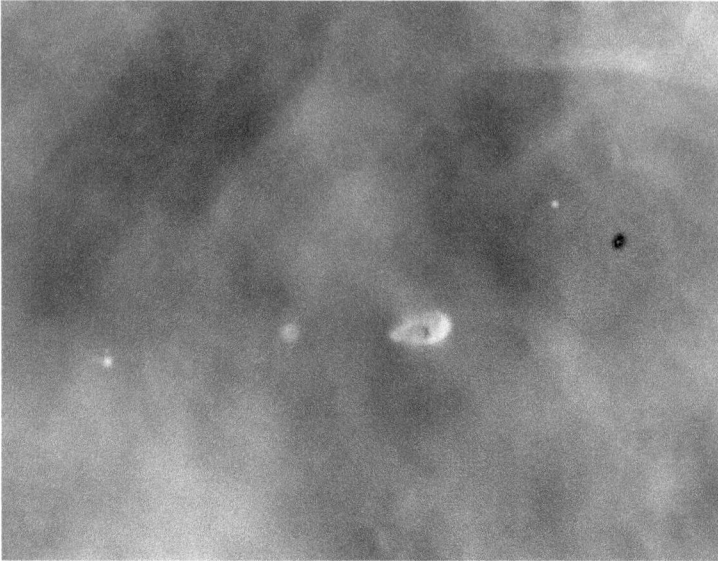

Figure 34: *Hubble image of protoplanetary discs in the Orion Nebula, a light-years-wide "stellar nursery" probably very similar to the primordial nebula from which the Sun formed*

just over that of the Sun ($M_\odot$) was about the same as that of the Sun today, with hydrogen, along with helium and trace amounts of lithium produced by Big Bang nucleosynthesis, forming about 98% of its mass. The remaining 2% of the mass consisted of heavier elements that were created by nucleosynthesis in earlier generations of stars.[300] Late in the life of these stars, they ejected heavier elements into the interstellar medium.

The oldest inclusions found in meteorites, thought to trace the first solid material to form in the pre-solar nebula, are 4568.2 million years old, which is one definition of the age of the Solar System. Studies of ancient meteorites reveal traces of stable daughter nuclei of short-lived isotopes, such as iron-60, that only form in exploding, short-lived stars. This indicates that one or more supernovae occurred near the Sun while it was forming. A shock wave from a supernova may have triggered the formation of the Sun by creating relatively dense regions within the cloud, causing these regions to collapse. Because only massive, short-lived stars produce supernovae, the Sun must have formed in a large star-forming region that produced massive stars, possibly similar to the Orion Nebula. Studies of the structure of the Kuiper belt and of anomalous materials within it suggest that the Sun formed within a cluster of between 1,000 and 10,000 stars with a diameter of between 6.5 and 19.5 light years and a collective mass of 3,000 $M_\odot$. This cluster began to break apart between 135 million and 535 million years after formation. Several simulations of our young Sun interacting with close-passing stars over the first 100 million years of its life produce anomalous orbits observed in the outer Solar System, such as detached objects.

Because of the conservation of angular momentum, the nebula spun faster as it collapsed. As the material within the nebula condensed, the atoms within it began to collide with increasing frequency, converting their kinetic energy into heat. The centre, where most of the mass collected, became increasingly hotter than the surrounding disc. Over about 100,000 years, the competing forces of gravity, gas pressure, magnetic fields, and rotation caused the contracting nebula to flatten into a spinning protoplanetary disc with a diameter of about 200 AU and form a hot, dense protostar (a star in which hydrogen fusion has not yet begun) at the centre.

At this point in its evolution, the Sun is thought to have been a T Tauri star. Studies of T Tauri stars show that they are often accompanied by discs of pre-planetary matter with masses of 0.001–0.1 $M_\odot$. These discs extend to several hundred AU—the Hubble Space Telescope has observed protoplanetary discs of up to 1000 AU in diameter in star-forming regions such as the Orion Nebula—and are rather cool, reaching a surface temperature of only about 1000 kelvin at their hottest. Within 50 million years, the temperature and pressure at the core of the Sun became so great that its hydrogen began to fuse, creating an internal source of energy that countered gravitational contraction until hydrostatic equilibrium was achieved. This marked the Sun's entry into the prime phase of its life, known as the main sequence. Main-sequence stars derive energy from the fusion of hydrogen into helium in their cores. The Sun remains a main-sequence star today.

Formation of the planets

The various planets are thought to have formed from the solar nebula, the disc-shaped cloud of gas and dust left over from the Sun's formation. The currently accepted method by which the planets formed is accretion, in which the planets began as dust grains in orbit around the central protostar. Through direct contact, these grains formed into clumps up to 200 metres in diameter, which in turn collided to form larger bodies (planetesimals) of ~10 kilometres (km) in size. These gradually increased through further collisions, growing at the rate of centimetres per year over the course of the next few million years.

The inner Solar System, the region of the Solar System inside 4 AU, was too warm for volatile molecules like water and methane to condense, so the planetesimals that formed there could only form from compounds with high

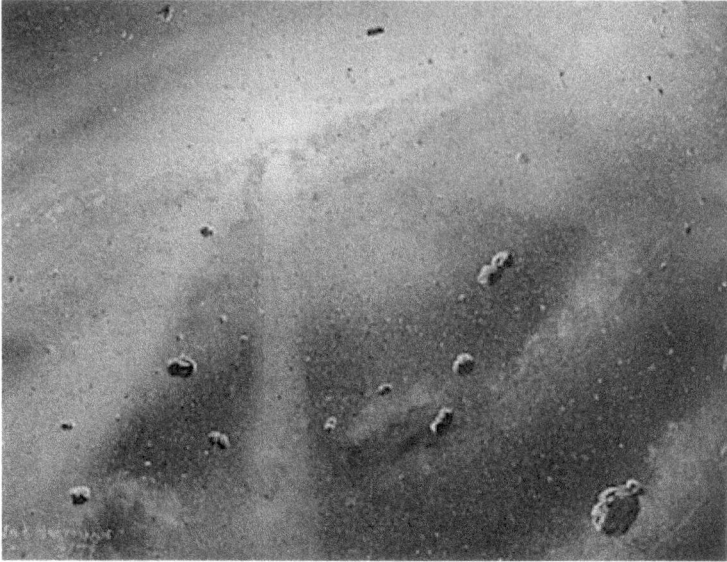

Figure 35: *Artist's conception of the solar nebula*

melting points, such as metals (like iron, nickel, and aluminium) and rocky silicates. These rocky bodies would become the terrestrial planets (Mercury, Venus, Earth, and Mars). These compounds are quite rare in the Universe, comprising only 0.6% of the mass of the nebula, so the terrestrial planets could not grow very large. The terrestrial embryos grew to about 0.05 Earth masses ($M_\oplus$) and ceased accumulating matter about 100,000 years after the formation of the Sun; subsequent collisions and mergers between these planet-sized bodies allowed terrestrial planets to grow to their present sizes (see Terrestrial planets below).

When the terrestrial planets were forming, they remained immersed in a disk of gas and dust. The gas was partially supported by pressure and so did not orbit the Sun as rapidly as the planets. The resulting drag and, more importantly, gravitational interactions with the surrounding material caused a transfer of angular momentum, and as a result the planets gradually migrated to new orbits. Models show that density and temperature variations in the disk governed this rate of migration, but the net trend was for the inner planets to migrate inward as the disk dissipated, leaving the planets in their current orbits.

The giant planets (Jupiter, Saturn, Uranus, and Neptune) formed further out, beyond the frost line, which is the point between the orbits of Mars and Jupiter where the material is cool enough for volatile icy compounds to remain solid.

The ices that formed the Jovian planets were more abundant than the metals and silicates that formed the terrestrial planets, allowing the giant planets to grow massive enough to capture hydrogen and helium, the lightest and most abundant elements. Planetesimals beyond the frost line accumulated up to 4 $M_\oplus$ within about 3 million years. Today, the four giant planets comprise just under 99% of all the mass orbiting the Sun.[301] Theorists believe it is no accident that Jupiter lies just beyond the frost line. Because the frost line accumulated large amounts of water via evaporation from infalling icy material, it created a region of lower pressure that increased the speed of orbiting dust particles and halted their motion toward the Sun. In effect, the frost line acted as a barrier that caused material to accumulate rapidly at $\sim$5 AU from the Sun. This excess material coalesced into a large embryo (or core) on the order of 10 $M_\oplus$, which began to accumulate an envelope via accretion of gas from the surrounding disc at an ever-increasing rate. Once the envelope mass became about equal to the solid core mass, growth proceeded very rapidly, reaching about 150 Earth masses $\sim$10^5 years thereafter and finally topping out at 318 $M_\oplus$. Saturn may owe its substantially lower mass simply to having formed a few million years after Jupiter, when there was less gas available to consume.

T Tauri stars like the young Sun have far stronger stellar winds than more stable, older stars. Uranus and Neptune are thought to have formed after Jupiter and Saturn did, when the strong solar wind had blown away much of the disc material. As a result, the planets accumulated little hydrogen and helium—not more than 1 $M_\oplus$ each. Uranus and Neptune are sometimes referred to as failed cores. The main problem with formation theories for these planets is the timescale of their formation. At the current locations it would have taken millions of years for their cores to accrete. This means that Uranus and Neptune may have formed closer to the Sun—near or even between Jupiter and Saturn—and later migrated or were ejected outward (see Planetary migration below). Motion in the planetesimal era was not all inward toward the Sun; the *Stardust* sample return from Comet Wild 2 has suggested that materials from the early formation of the Solar System migrated from the warmer inner Solar System to the region of the Kuiper belt.

After between three and ten million years, the young Sun's solar wind would have cleared away all the gas and dust in the protoplanetary disc, blowing it into interstellar space, thus ending the growth of the planets.

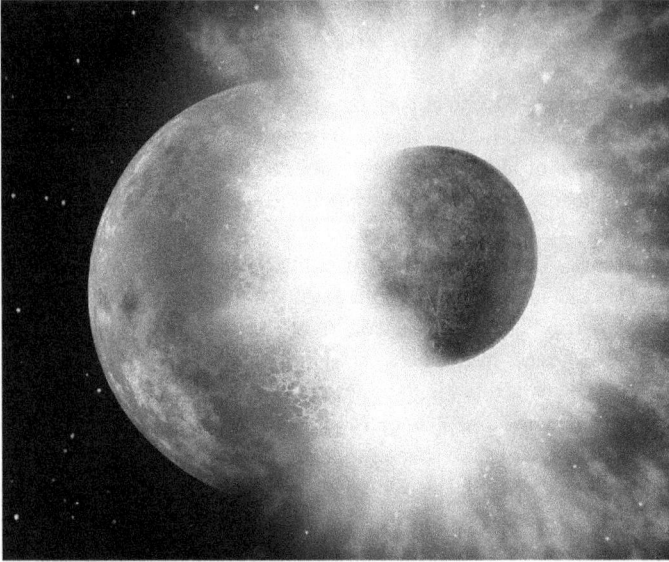

Figure 36: *Artist's conception of the giant impact thought to have formed the Moon*

Subsequent evolution

The planets were originally thought to have formed in or near their current orbits. From that a minimum mass of the nebula i.e. the protoplanetary disc, was derived that was necessary to form the planets - the minimum mass solar nebula. It was derived that the nebula mass must have exceeded 3585 times that of the Earth.

However, this has been questioned during the last 20 years. Currently, many planetary scientists think that the Solar System might have looked very different after its initial formation: several objects at least as massive as Mercury were present in the inner Solar System, the outer Solar System was much more compact than it is now, and the Kuiper belt was much closer to the Sun.

Terrestrial planets

At the end of the planetary formation epoch the inner Solar System was populated by 50–100 Moon- to Mars-sized planetary embryos. Further growth was possible only because these bodies collided and merged, which took less than 100 million years. These objects would have gravitationally interacted with one another, tugging at each other's orbits until they collided, growing larger until the four terrestrial planets we know today took shape. One such giant collision is thought to have formed the Moon (see Moons below), while another removed the outer envelope of the young Mercury.

One unresolved issue with this model is that it cannot explain how the initial orbits of the proto-terrestrial planets, which would have needed to be highly eccentric to collide, produced the remarkably stable and nearly circular orbits they have today. One hypothesis for this "eccentricity dumping" is that the terrestrials formed in a disc of gas still not expelled by the Sun. The "gravitational drag" of this residual gas would have eventually lowered the planets' energy, smoothing out their orbits. However, such gas, if it existed, would have prevented the terrestrial planets' orbits from becoming so eccentric in the first place. Another hypothesis is that gravitational drag occurred not between the planets and residual gas but between the planets and the remaining small bodies. As the large bodies moved through the crowd of smaller objects, the smaller objects, attracted by the larger planets' gravity, formed a region of higher density, a "gravitational wake", in the larger objects' path. As they did so, the increased gravity of the wake slowed the larger objects down into more regular orbits.

Asteroid belt

The outer edge of the terrestrial region, between 2 and 4 AU from the Sun, is called the asteroid belt. The asteroid belt initially contained more than enough matter to form 2–3 Earth-like planets, and, indeed, a large number of planetesimals formed there. As with the terrestrials, planetesimals in this region later coalesced and formed 20–30 Moon- to Mars-sized planetary embryos; however, the proximity of Jupiter meant that after this planet formed, 3 million years after the Sun, the region's history changed dramatically. Orbital resonances with Jupiter and Saturn are particularly strong in the asteroid belt, and gravitational interactions with more massive embryos scattered many planetesimals into those resonances. Jupiter's gravity increased the velocity of objects within these resonances, causing them to shatter upon collision with other bodies, rather than accrete.

As Jupiter migrated inward following its formation (see Planetary migration below), resonances would have swept across the asteroid belt, dynamically exciting the region's population and increasing their velocities relative to each other. The cumulative action of the resonances and the embryos either scattered the planetesimals away from the asteroid belt or excited their orbital inclinations and eccentricities. Some of those massive embryos too were ejected by Jupiter, while others may have migrated to the inner Solar System and played a role in the final accretion of the terrestrial planets. During this primary depletion period, the effects of the giant planets and planetary embryos left the

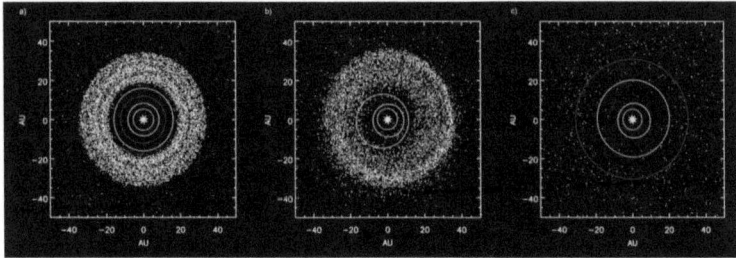

Figure 37: *Simulation showing outer planets and Kuiper belt: a) Before Jupiter/ Saturn 2:1 resonance b) Scattering of Kuiper belt objects into the Solar System after the orbital shift of Neptune c) After ejection of Kuiper belt bodies by Jupiter*

asteroid belt with a total mass equivalent to less than 1% that of the Earth, composed mainly of small planetesimals. This is still 10–20 times more than the current mass in the main belt, which is now about $1/2,000$ $M_\oplus$. A secondary depletion period that brought the asteroid belt down close to its present mass is thought to have followed when Jupiter and Saturn entered a temporary 2:1 orbital resonance (see below).

The inner Solar System's period of giant impacts probably played a role in the Earth acquiring its current water content ($\sim 6 \times 10^{21}$ kg) from the early asteroid belt. Water is too volatile to have been present at Earth's formation and must have been subsequently delivered from outer, colder parts of the Solar System. The water was probably delivered by planetary embryos and small planetesimals thrown out of the asteroid belt by Jupiter. A population of main-belt comets discovered in 2006 has been also suggested as a possible source for Earth's water. In contrast, comets from the Kuiper belt or farther regions delivered not more than about 6% of Earth's water. The panspermia hypothesis holds that life itself may have been deposited on Earth in this way, although this idea is not widely accepted.

Planetary migration

According to the nebular hypothesis, the outer two planets may be in the "wrong place". Uranus and Neptune (known as the "ice giants") exist in a region where the reduced density of the solar nebula and longer orbital times render their formation highly implausible. The two are instead thought to have formed in orbits near Jupiter and Saturn, where more material was available, and to have migrated outward to their current positions over hundreds of millions of years.

The migration of the outer planets is also necessary to account for the existence and properties of the Solar System's outermost regions. Beyond Neptune, the

Solar System continues into the Kuiper belt, the scattered disc, and the Oort cloud, three sparse populations of small icy bodies thought to be the points of origin for most observed comets. At their distance from the Sun, accretion was too slow to allow planets to form before the solar nebula dispersed, and thus the initial disc lacked enough mass density to consolidate into a planet. The Kuiper belt lies between 30 and 55 AU from the Sun, while the farther scattered disc extends to over 100 AU, and the distant Oort cloud begins at about 50,000 AU. Originally, however, the Kuiper belt was much denser and closer to the Sun, with an outer edge at approximately 30 AU. Its inner edge would have been just beyond the orbits of Uranus and Neptune, which were in turn far closer to the Sun when they formed (most likely in the range of 15–20 AU), and in 50% of simulations ended up opposite locations, with Uranus farther from the Sun than Neptune.

According to the Nice model, after the formation of the Solar System, the orbits of all the giant planets continued to change slowly, influenced by their interaction with the large number of remaining planetesimals. After 500–600 million years (about 4 billion years ago) Jupiter and Saturn fell into a 2:1 resonance: Saturn orbited the Sun once for every two Jupiter orbits. This resonance created a gravitational push against the outer planets, possibly causing Neptune to surge past Uranus and plough into the ancient Kuiper belt. The planets scattered the majority of the small icy bodies inwards, while themselves moving outwards. These planetesimals then scattered off the next planet they encountered in a similar manner, moving the planets' orbits outwards while they moved inwards. This process continued until the planetesimals interacted with Jupiter, whose immense gravity sent them into highly elliptical orbits or even ejected them outright from the Solar System. This caused Jupiter to move slightly inward.[302] Those objects scattered by Jupiter into highly elliptical orbits formed the Oort cloud; those objects scattered to a lesser degree by the migrating Neptune formed the current Kuiper belt and scattered disc. This scenario explains the Kuiper belt's and scattered disc's present low mass. Some of the scattered objects, including Pluto, became gravitationally tied to Neptune's orbit, forcing them into mean-motion resonances. Eventually, friction within the planetesimal disc made the orbits of Uranus and Neptune circular again.

In contrast to the outer planets, the inner planets are not thought to have migrated significantly over the age of the Solar System, because their orbits have remained stable following the period of giant impacts.

Another question is why Mars came out so small compared with Earth. A study by Southwest Research Institute, San Antonio, Texas, published June 6, 2011 (called the Grand tack hypothesis), proposes that Jupiter had migrated inward to 1.5 AU. After Saturn formed, migrated inward, and established the 2:3 mean motion resonance with Jupiter, the study assumes that both planets migrated back to their present positions. Jupiter thus would have consumed much of the material that would have created a bigger Mars. The same simulations also reproduce the characteristics of the modern asteroid belt, with dry asteroids and water-rich objects similar to comets. However, it is unclear whether conditions in the solar nebula would have allowed Jupiter and Saturn to move back to their current positions, and according to current estimates this possibility appears unlikely. Moreover, alternative explanations for the small mass of Mars exist.

Late Heavy Bombardment and after

Life timeline

θ —

500
1000
1500
2000
2500
3000
3500
4000
4500

Figure 38: *Meteor Crater in Arizona. Created 50,000
years ago by an impactor about 50 metres (160 ft) across,
it shows that the accretion of the Solar System is not over.*

Axis scale: million years

Also see: *Human timeline* and *Nature timeline*

Gravitational disruption from the outer planets' migration would have sent
large numbers of asteroids into the inner Solar System, severely depleting the
original belt until it reached today's extremely low mass. This event may have
triggered the Late Heavy Bombardment that occurred approximately 4 billion
years ago, 500–600 million years after the formation of the Solar System. This
period of heavy bombardment lasted several hundred million years and is evi-
dent in the cratering still visible on geologically dead bodies of the inner Solar
System such as the Moon and Mercury. The oldest known evidence for life on
Earth dates to 3.8 billion years ago—almost immediately after the end of the
Late Heavy Bombardment.

Impacts are thought to be a regular (if currently infrequent) part of the evolution of the Solar System. That they continue to happen is evidenced by the
collision of Comet Shoemaker–Levy 9 with Jupiter in 1994, the 2009 Jupiter
impact event, the Tunguska event, the Chelyabinsk meteor and the impact feature Meteor Crater in Arizona. The process of accretion, therefore, is not
complete, and may still pose a threat to life on Earth.

Over the course of the Solar System's evolution, comets were ejected out of the inner Solar System by the gravity of the giant planets, and sent thousands of AU outward to form the Oort cloud, a spherical outer swarm of cometary nuclei at the farthest extent of the Sun's gravitational pull. Eventually, after about 800 million years, the gravitational disruption caused by galactic tides, passing stars and giant molecular clouds began to deplete the cloud, sending comets into the inner Solar System. The evolution of the outer Solar System also appears to have been influenced by space weathering from the solar wind, micrometeorites, and the neutral components of the interstellar medium.

The evolution of the asteroid belt after Late Heavy Bombardment was mainly governed by collisions. Objects with large mass have enough gravity to retain any material ejected by a violent collision. In the asteroid belt this usually is not the case. As a result, many larger objects have been broken apart, and sometimes newer objects have been forged from the remnants in less violent collisions. Moons around some asteroids currently can only be explained as consolidations of material flung away from the parent object without enough energy to entirely escape its gravity.

Moons

Moons have come to exist around most planets and many other Solar System bodies. These natural satellites originated by one of three possible mechanisms:

- Co-formation from a circumplanetary disc (only in the cases of the giant planets);
- Formation from impact debris (given a large enough impact at a shallow angle); and
- Capture of a passing object.

Jupiter and Saturn have several large moons, such as Io, Europa, Ganymede and Titan, which may have originated from discs around each giant planet in much the same way that the planets formed from the disc around the Sun.[303] This origin is indicated by the large sizes of the moons and their proximity to the planet. These attributes are impossible to achieve via capture, while the gaseous nature of the primaries also make formation from collision debris unlikely. The outer moons of the giant planets tend to be small and have eccentric orbits with arbitrary inclinations. These are the characteristics expected of captured bodies. Most such moons orbit in the direction opposite the rotation of their primary. The largest irregular moon is Neptune's moon Triton, which is thought to be a captured Kuiper belt object.

Moons of solid Solar System bodies have been created by both collisions and capture. Mars's two small moons, Deimos and Phobos, are thought to be

captured asteroids.[304] The Earth's Moon is thought to have formed as a result of a single, large head-on collision. The impacting object probably had a
mass comparable to that of Mars, and the impact probably occurred near the
end of the period of giant impacts. The collision kicked into orbit some of
the impactor's mantle, which then coalesced into the Moon. The impact was
probably the last in the series of mergers that formed the Earth. It has been
further hypothesized that the Mars-sized object may have formed at one of the
stable Earth–Sun Lagrangian points (either L_4 or L_5) and drifted from its position. The moons of trans-Neptunian objects Pluto (Charon) and Orcus (Vanth)
may also have formed by means of a large collision: the Pluto–Charon, Orcus–Vanth and Earth–Moon systems are unusual in the Solar System in that
the satellite's mass is at least 1% that of the larger body.

Future

Astronomers estimate that the Solar System as we know it today will not
change drastically until the Sun has fused almost all the hydrogen fuel in
its core into helium, beginning its evolution from the main sequence of the
Hertzsprung–Russell diagram and into its red-giant phase. Even so, the Solar
System will continue to evolve until then.

Long-term stability

The Solar System is chaotic over million- and billion-year timescales, with the
orbits of the planets open to long-term variations. One notable example of
this chaos is the Neptune–Pluto system, which lies in a 3:2 orbital resonance.
Although the resonance itself will remain stable, it becomes impossible to predict the position of Pluto with any degree of accuracy more than 10–20 million
years (the Lyapunov time) into the future. Another example is Earth's axial
tilt, which, due to friction raised within Earth's mantle by tidal interactions
with the Moon (see below), will be incomputable at some point between 1.5
and 4.5 billion years from now.

The outer planets' orbits are chaotic over longer timescales, with a Lyapunov
time in the range of 2–230 million years. In all cases this means that the position of a planet along its orbit ultimately becomes impossible to predict with
any certainty (so, for example, the timing of winter and summer become uncer-
tain), but in some cases the orbits themselves may change dramatically. Such
chaos manifests most strongly as changes in eccentricity, with some planets'
orbits becoming significantly more—or less—elliptical.

Ultimately, the Solar System is stable in that none of the planets are likely
to collide with each other or be ejected from the system in the next few billion years. Beyond this, within five billion years or so Mars's eccentricity may

grow to around 0.2, such that it lies on an Earth-crossing orbit, leading to a potential collision. In the same timescale, Mercury's eccentricity may grow even further, and a close encounter with Venus could theoretically eject it from the Solar System altogether or send it on a collision course with Venus or Earth. This could happen within a billion years, according to numerical simulations in which Mercury's orbit is perturbed.

Moon–ring systems

The evolution of moon systems is driven by tidal forces. A moon will raise a tidal bulge in the object it orbits (the primary) due to the differential gravitational force across diameter of the primary. If a moon is revolving in the same direction as the planet's rotation and the planet is rotating faster than the orbital period of the moon, the bulge will constantly be pulled ahead of the moon. In this situation, angular momentum is transferred from the rotation of the primary to the revolution of the satellite. The moon gains energy and gradually spirals outward, while the primary rotates more slowly over time.

The Earth and its Moon are one example of this configuration. Today, the Moon is tidally locked to the Earth; one of its revolutions around the Earth (currently about 29 days) is equal to one of its rotations about its axis, so it always shows one face to the Earth. The Moon will continue to recede from Earth, and Earth's spin will continue to slow gradually. In about 50 billion years, if they survive the Sun's expansion, the Earth and Moon will become tidally locked to each other; each will be caught up in what is called a "spin–orbit resonance" in which the Moon will circle the Earth in about 47 days and both Moon and Earth will rotate around their axes in the same time, each only visible from one hemisphere of the other. Other examples are the Galilean moons of Jupiter (as well as many of Jupiter's smaller moons) and most of the larger moons of Saturn.

A different scenario occurs when the moon is either revolving around the primary faster than the primary rotates, or is revolving in the direction opposite the planet's rotation. In these cases, the tidal bulge lags behind the moon in its orbit. In the former case, the direction of angular momentum transfer is reversed, so the rotation of the primary speeds up while the satellite's orbit shrinks. In the latter case, the angular momentum of the rotation and revolution have opposite signs, so transfer leads to decreases in the magnitude of each (that cancel each other out).[305] In both cases, tidal deceleration causes the moon to spiral in towards the primary until it either is torn apart by tidal stresses, potentially creating a planetary ring system, or crashes into the planet's surface or atmosphere. Such a fate awaits the moons Phobos of Mars (within 30 to 50 million years), Triton of Neptune (in 3.6 billion years), Metis and

Figure 39: *Neptune and its moon Triton, taken by Voyager 2. Triton's orbit will eventually take it within Neptune's Roche limit, tearing it apart and possibly forming a new ring system.*

Adrastea of Jupiter, and at least 16 small satellites of Uranus and Neptune. Uranus's Desdemona may even collide with one of its neighboring moons.[306]

A third possibility is where the primary and moon are tidally locked to each other. In that case, the tidal bulge stays directly under the moon, there is no transfer of angular momentum, and the orbital period will not change. Pluto and Charon are an example of this type of configuration.

Prior to the 2004 arrival of the *Cassini–Huygens* spacecraft, the rings of Saturn were widely thought to be much younger than the Solar System and were not expected to survive beyond another 300 million years. Gravitational interactions with Saturn's moons were expected to gradually sweep the rings' outer edge toward the planet, with abrasion by meteorites and Saturn's gravity eventually taking the rest, leaving Saturn unadorned. However, data from the *Cassini* mission led scientists to revise that early view. Observations revealed 10 km-wide icy clumps of material that repeatedly break apart and reform, keeping the rings fresh. Saturn's rings are far more massive than the rings of the other giant planets. This large mass is thought to have preserved Saturn's rings since it first formed 4.5 billion years ago, and is likely to preserve them for billions of years to come.

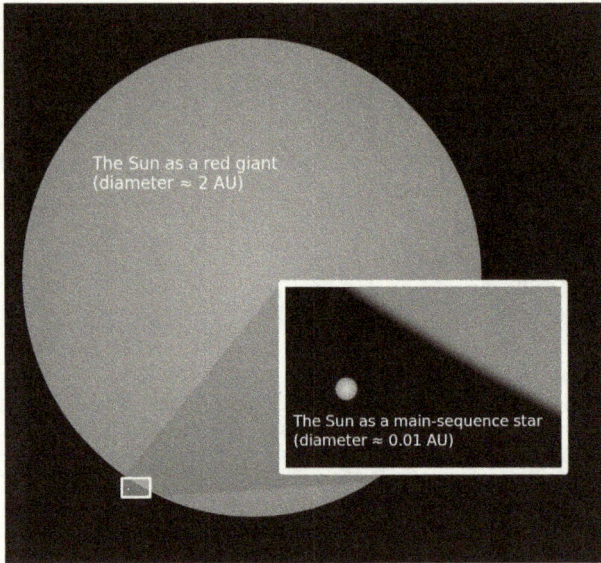

Figure 40: *Relative size of the Sun as it is now (inset) compared to its estimated future size as a red giant*

The Sun and planetary environments

In the long term, the greatest changes in the Solar System will come from changes in the Sun itself as it ages. As the Sun burns through its supply of hydrogen fuel, it gets hotter and burns the remaining fuel even faster. As a result, the Sun is growing brighter at a rate of ten percent every 1.1 billion years. In one billion years' time, as the Sun's radiation output increases, its circumstellar habitable zone will move outwards, making the Earth's surface too hot for liquid water to exist there naturally. At this point, all life on land will become extinct. Evaporation of water, a potent greenhouse gas, from the oceans' surface could accelerate temperature increase, potentially ending all life on Earth even sooner. During this time, it is possible that as Mars's surface temperature gradually rises, carbon dioxide and water currently frozen under the surface regolith will release into the atmosphere, creating a greenhouse effect that will heat the planet until it achieves conditions parallel to Earth today, providing a potential future abode for life. By 3.5 billion years from now, Earth's surface conditions will be similar to those of Venus today.

Around 5.4 billion years from now, the core of the Sun will become hot enough to trigger hydrogen fusion in its surrounding shell. This will cause the outer layers of the star to expand greatly, and the star will enter a phase of its life in which it is called a red giant.[307] Within 7.5 billion years, the Sun will have expanded to a radius of 1.2 AU—256 times its current size. At the tip of the red giant branch, as a result of the vastly increased surface area, the Sun's surface will be much cooler (about 2600 K) than now and its luminosity much higher—up to 2,700 current solar luminosities. For part of its red giant life, the Sun will have a strong stellar wind that will carry away around 33% of its mass.[308] During these times, it is possible that Saturn's moon Titan could achieve surface temperatures necessary to support life.

As the Sun expands, it will swallow the planets Mercury and Venus. Earth's fate is less clear; although the Sun will envelop Earth's current orbit, the star's loss of mass (and thus weaker gravity) will cause the planets' orbits to move farther out. If it were only for this, Venus and Earth would probably escape incineration, but a 2008 study suggests that Earth will likely be swallowed up as a result of tidal interactions with the Sun's weakly bound outer envelope.

Gradually, the hydrogen burning in the shell around the solar core will increase the mass of the core until it reaches about 45% of the present solar mass. At this point the density and temperature will become so high that the fusion of helium into carbon will begin, leading to a helium flash; the Sun will shrink from around 250 to 11 times its present (main-sequence) radius. Consequently, its luminosity will decrease from around 3,000 to 54 times its current level, and its surface temperature will increase to about 4770 K. The Sun will become a horizontal giant, burning helium in its core in a stable fashion much like it burns hydrogen today. The helium-fusing stage will last only 100 million years. Eventually, it will have to again resort to the reserves of hydrogen and helium in its outer layers and will expand a second time, turning into what is known as an asymptotic giant. Here the luminosity of the Sun will increase again, reaching about 2,090 present luminosities, and it will cool to about 3500 K. This phase lasts about 30 million years, after which, over the course of a further 100,000 years, the Sun's remaining outer layers will fall away, ejecting a vast stream of matter into space and forming a halo known (misleadingly) as a planetary nebula. The ejected material will contain the helium and carbon produced by the Sun's nuclear reactions, continuing the enrichment of the interstellar medium with heavy elements for future generations of stars.

This is a relatively peaceful event, nothing akin to a supernova, which the Sun is too small to undergo as part of its evolution. Any observer present to witness this occurrence would see a massive increase in the speed of the solar wind, but not enough to destroy a planet completely. However, the star's loss of mass could send the orbits of the surviving planets into chaos, causing some to collide, others to be ejected from the Solar System, and still others to be torn apart by tidal interactions. Afterwards, all that will remain of the Sun is

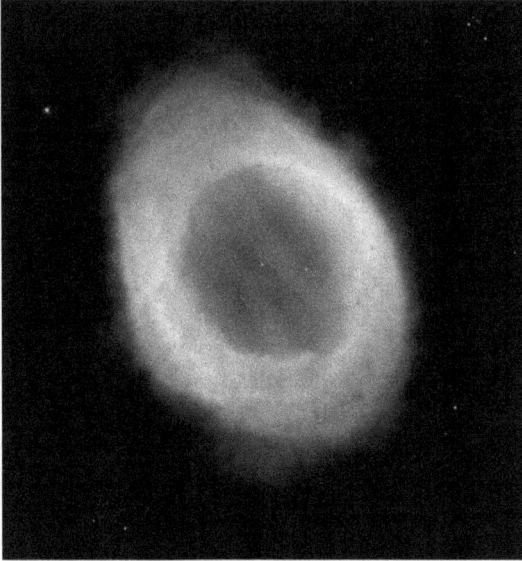

Figure 41: *The Ring nebula, a planetary nebula similar to what the Sun will become*

a white dwarf, an extraordinarily dense object, 54% its original mass but only
the size of the Earth. Initially, this white dwarf may be 100 times as luminous
as the Sun is now. It will consist entirely of degenerate carbon and oxygen,
but will never reach temperatures hot enough to fuse these elements. Thus the
white dwarf Sun will gradually cool, growing dimmer and dimmer.

As the Sun dies, its gravitational pull on the orbiting bodies such as planets,
comets and asteroids will weaken due to its mass loss. All remaining planets'
orbits will expand; if Venus, Earth, and Mars still exist, their orbits will lie
roughly at 1.4 AU (210,000,000 km), 1.9 AU (280,000,000 km), and 2.8 AU
(420,000,000 km). They and the other remaining planets will become dark,
frigid hulks, completely devoid of any form of life. They will continue to orbit
their star, their speed slowed due to their increased distance from the Sun and
the Sun's reduced gravity. Two billion years later, when the Sun has cooled to
the 6000–8000K range, the carbon and oxygen in the Sun's core will freeze,
with over 90% of its remaining mass assuming a crystalline structure. Eventually, after billions more years, the Sun will finally cease to shine altogether,
becoming a black dwarf.

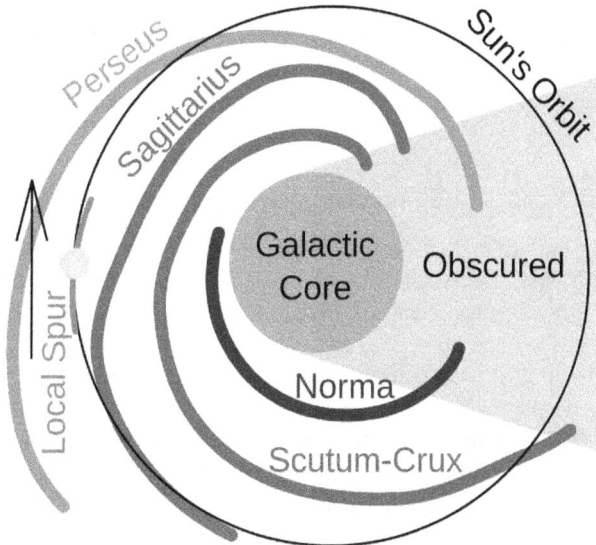

Figure 42: *Location of the Solar System within the Milky Way*

Galactic interaction

The Solar System travels alone through the Milky Way in a circular orbit approximately 30,000 light years from the Galactic Centre. Its speed is about 220 km/s. The period required for the Solar System to complete one revolution around the Galactic Centre, the galactic year, is in the range of 220–250 million years. Since its formation, the Solar System has completed at least 20 such revolutions.

Various scientists have speculated that the Solar System's path through the galaxy is a factor in the periodicity of mass extinctions observed in the Earth's fossil record. One hypothesis supposes that vertical oscillations made by the Sun as it orbits the Galactic Centre cause it to regularly pass through the galactic plane. When the Sun's orbit takes it outside the galactic disc, the influence of the galactic tide is weaker; as it re-enters the galactic disc, as it does every 20–25 million years, it comes under the influence of the far stronger "disc tides", which, according to mathematical models, increase the flux of Oort cloud comets into the Solar System by a factor of 4, leading to a massive increase in the likelihood of a devastating impact.

However, others argue that the Sun is currently close to the galactic plane, and yet the last great extinction event was 15 million years ago. Therefore, the Sun's vertical position cannot alone explain such periodic extinctions, and that extinctions instead occur when the Sun passes through the galaxy's spiral arms. Spiral arms are home not only to larger numbers of molecular clouds, whose gravity may distort the Oort cloud, but also to higher concentrations of bright blue giants, which live for relatively short periods and then explode violently as supernovae.

Galactic collision and planetary disruption

Although the vast majority of galaxies in the Universe are moving away from the Milky Way, the Andromeda Galaxy, the largest member of the Local Group of galaxies, is heading toward it at about 120 km/s. In 4 billion years, Andromeda and the Milky Way will collide, causing both to deform as tidal forces distort their outer arms into vast tidal tails. If this initial disruption occurs, astronomers calculate a 12% chance that the Solar System will be pulled outward into the Milky Way's tidal tail and a 3% chance that it will become gravitationally bound to Andromeda and thus a part of that galaxy. After a further series of glancing blows, during which the likelihood of the Solar System's ejection rises to 30%, the galaxies' supermassive black holes will merge. Eventually, in roughly 6 billion years, the Milky Way and Andromeda will complete their merger into a giant elliptical galaxy. During the merger, if there is enough gas, the increased gravity will force the gas to the centre of the forming elliptical galaxy. This may lead to a short period of intensive star formation called a starburst. In addition, the infalling gas will feed the newly formed black hole, transforming it into an active galactic nucleus. The force of these interactions will likely push the Solar System into the new galaxy's outer halo, leaving it relatively unscathed by the radiation from these collisions.

It is a common misconception that this collision will disrupt the orbits of the planets in the Solar System. Although it is true that the gravity of passing stars can detach planets into interstellar space, distances between stars are so great that the likelihood of the Milky Way–Andromeda collision causing such disruption to any individual star system is negligible. Although the Solar System as a whole could be affected by these events, the Sun and planets are not expected to be disturbed.

However, over time, the cumulative probability of a chance encounter with a star increases, and disruption of the planets becomes all but inevitable. Assuming that the Big Crunch or Big Rip scenarios for the end of the Universe do not occur, calculations suggest that the gravity of passing stars will have completely stripped the dead Sun of its remaining planets within 1 quadrillion (10^{15}) years. This point marks the end of the Solar System. Although the Sun and planets may survive, the Solar System, in any meaningful sense, will cease to exist.

Chronology

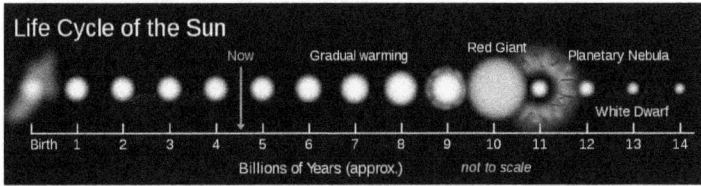

The time frame of the Solar System's formation has been determined using radiometric dating. Scientists estimate that the Solar System is 4.6 billion years old. The oldest known mineral grains on Earth are approximately 4.4 billion years old. Rocks this old are rare, as Earth's surface is constantly being reshaped by erosion, volcanism, and plate tectonics. To estimate the age of the Solar System, scientists use meteorites, which were formed during the early condensation of the solar nebula. Almost all meteorites (see the Canyon Diablo meteorite) are found to have an age of 4.6 billion years, suggesting that the Solar System must be at least this old.

Studies of discs around other stars have also done much to establish a time frame for Solar System formation. Stars between one and three million years old have discs rich in gas, whereas discs around stars more than 10 million years old have little to no gas, suggesting that giant planets within them have ceased forming.

Timeline of Solar System evolution

A graphical timeline is available at
Graphical timeline of Earth and Sun

Note: All dates and times in this chronology are approximate and should be taken as an order of magnitude indicator only.

Chronology of the formation and evolution of the Solar System

Phase	Time since formation of the Sun	Time from present (approximate)	Event
Pre-Solar System	Billions of years before the formation of the Solar System	Over 4.6 billion years ago (bya)	Previous generations of stars live and die, injecting heavy elements into the interstellar medium out of which the Solar System formed.

	~ 50 million years before formation of the Solar System	4.6 bya	If the Solar System formed in an Orion nebula-like star-forming region, the most massive stars are formed, live their lives, die, and explode in supernova. One particular supernova, called the *primal supernova*, possibly triggers the formation of the Solar System.
Formation of Sun	0–100,000 years	4.6 bya	Pre-solar nebula forms and begins to collapse. Sun begins to form.
	100,000 – 50 million years	4.6 bya	Sun is a T Tauri protostar.
	100,000 – 10 million years	4.6 bya	By 10 million years, gas in the protoplanetary disc has been blown away, and outer planet formation is likely complete.
	10 million – 100 million years	4.5–4.6 bya	Terrestrial planets and the Moon form. Giant impacts occur. Water delivered to Earth.
Main sequence	50 million years	4.5 bya	Sun becomes a main-sequence star.
	200 million years	4.4 bya	Oldest known rocks on the Earth formed.
	500 million – 600 million years	4.0–4.1 bya	Resonance in Jupiter and Saturn's orbits moves Neptune out into the Kuiper belt. Late Heavy Bombardment occurs in the inner Solar System.
	800 million years	3.8 bya	Oldest known life on Earth. Oort cloud reaches maximum mass.
	4.6 billion years	**Today**	Sun remains a main-sequence star, continually growing warmer and brighter by ~10% every 1 billion years.
	6 billion years	1.4 billion years in the future	Sun's habitable zone moves outside of the Earth's orbit, possibly shifting onto Mars's orbit.
	7 billion years	2.4 billion years in the future	The Milky Way and Andromeda Galaxy begin to collide. Slight chance the Solar System could be captured by Andromeda before the two galaxies fuse completely.
Post–main sequence	10 billion – 12 billion years	5–7 billion years in the future	Sun starts burning hydrogen in a shell surrounding its core, ending its main sequence life. Sun begins to ascend the red giant branch of the Hertzsprung–Russell diagram, growing dramatically more luminous (by a factor of up to 2,700), larger (by a factor of up to 250 in radius), and cooler (down to 2600 K): Sun is now a red giant. Mercury and possibly Venus and Earth are swallowed. Saturn's moon Titan may become habitable.
	~ 12 billion years	~ 7 billion years in the future	Sun passes through helium-burning horizontal-branch and asymptotic-giant-branch phases, losing a total of ~30% of its mass in all post-main-sequence phases. The asymptotic-giant-branch phase ends with the ejection of a planetary nebula, leaving the core of the Sun behind as a white dwarf.

Remnant Sun	~ 1 quadrillion years (10^{15} years)	~ 1 quadrillion years in the future	Sun cools to 5 K. Gravity of passing stars detaches planets from orbits. Solar System ceases to exist.

Bibliography

- Duncan, Martin J.; Lissauer, Jack J. (1997). "Orbital Stability of the Uranian Satellite System". *Icarus*. **125** (1): 1–12. Bibcode: 1997Icar.. 125....1D[309]. doi: 10.1006/icar.1996.5568[310].
- Zeilik, Michael A.; Gregory, Stephen A. (1998). *Introductory Astronomy & Astrophysics* (4th ed.). Saunders College Publishing. ISBN 0-03-006228-4.

External links

- 7M animation[311] from skyandtelescope.com[312] showing the early evolution of the outer Solar System.
- QuickTime animation of the future collision between the Milky Way and Andromeda[313]
- How the Sun Will Die: And What Happens to Earth[314] (Video at Space.com)

<indicator name="featured-star"> ⭐ </indicator>

Moons

Moons of Uranus

Uranus is the seventh planet of the Solar System; it has 27 known moons, all of which are named after characters from the works of William Shakespeare and Alexander Pope. Uranus's moons are divided into three groups: thirteen inner moons, five major moons, and nine irregular moons. The inner moons are small dark bodies that share common properties and origins with Uranus's rings. The five major moons are massive enough to have reached hydrostatic equilibrium, and four of them show signs of internally driven processes such as canyon formation and volcanism on their surfaces. The largest of these five, Titania, is 1,578 km in diameter and the eighth-largest moon in the Solar System, and about one-twentieth the mass the Earth's Moon. The orbits of the regular moons are nearly coplanar with Uranus's equator, which is tilted 97.77° to its orbit. Uranus's irregular moons have elliptical and strongly inclined (mostly retrograde) orbits at large distances from the planet.

William Herschel discovered the first two moons, Titania and Oberon, in 1787, and the other three ellipsoidal moons were discovered in 1851 by William Lassell (Ariel and Umbriel) and in 1948 by Gerard Kuiper (Miranda). These five have planetary mass, and so would be considered (dwarf) planets if they were in direct orbit about the Sun. The remaining moons were discovered after 1985, either during the *Voyager 2* flyby mission or with the aid of advanced Earth-based telescopes.

Discovery

The first two moons to be discovered were Titania and Oberon, which were spotted by Sir William Herschel on January 11, 1787, six years after he had discovered the planet itself. Later, Herschel thought he had discovered up to six moons (see below) and perhaps even a ring. For nearly 50 years, Herschel's instrument was the only one with which the moons had been seen. In the 1840s,

Figure 43: *Uranus and its six largest moons compared at their proper relative sizes and relative positions. From left to right: Puck, Miranda, Ariel, Umbriel, Titania, and Oberon*

better instruments and a more favorable position of Uranus in the sky led to sporadic indications of satellites additional to Titania and Oberon. Eventually, the next two moons, Ariel and Umbriel, were discovered by William Lassell in 1851. The Roman numbering scheme of Uranus's moons was in a state of flux for a considerable time, and publications hesitated between Herschel's designations (where Titania and Oberon are Uranus II and IV) and William Lassell's (where they are sometimes I and II). With the confirmation of Ariel and Umbriel, Lassell numbered the moons I through IV from Uranus outward, and this finally stuck. In 1852, Herschel's son John Herschel gave the four then-known moons their names.

No other discoveries were made for almost another century. In 1948, Gerard Kuiper at the McDonald Observatory discovered the smallest and the last of the five large, spherical moons, Miranda. Decades later, the flyby of the *Voyager 2* space probe in January 1986 led to the discovery of ten further inner moons. Another satellite, Perdita, was discovered in 1999 after studying old *Voyager* photographs.

Uranus was the last giant planet without any known irregular moons, but since 1997 nine distant irregular moons have been identified using ground-based telescopes. Two more small inner moons, Cupid and Mab, were discovered

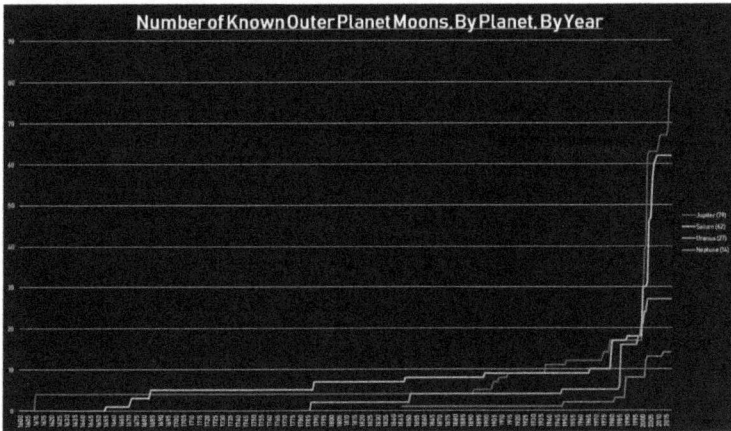

Figure 44: *The number of moons known for each of the four outer planets up to July 2018. Uranus currently has 27 known satellites.*

using the Hubble Space Telescope in 2003. As of 2016, the moon Margaret was the last Uranian moon discovered, and its characteristics were published in October 2003.

Spurious moons

After Herschel discovered Titania and Oberon on January 11, 1787, he subsequently believed that he had observed four other moons: two on January 18 and February 9, 1790, and two more on February 28 and March 26, 1794. It was thus believed for many decades thereafter that Uranus had a system of six satellites, though the four latter moons were never confirmed by any other astronomer. Lassell's observations of 1851, in which he discovered Ariel and Umbriel, however, failed to support Herschel's observations; Ariel and Umbriel, which Herschel certainly ought to have seen if he had seen any satellites beside Titania and Oberon, did not correspond to any of Herschel's four additional satellites in orbital characteristics. Herschel's four spurious satellites were thought to have sidereal periods of 5.89 days (interior to Titania), 10.96 days (between Titania and Oberon), 38.08 days, and 107.69 days (exterior to Oberon). It was therefore concluded that Herschel's four satellites were spurious, probably arising from the misidentification of faint stars in the vicinity of Uranus as satellites, and the credit for the discovery of Ariel and Umbriel was given to Lassell.

Names

Although the first two Uranian moons were discovered in 1787, they were not named until 1852, a year after two more moons had been discovered. The responsibility for naming was taken by John Herschel, son of the discoverer of Uranus. Herschel, instead of assigning names from Greek mythology, named the moons after magical spirits in English literature: the fairies Oberon and Titania from William Shakespeare's *A Midsummer Night's Dream*, and the sylphs Ariel and Umbriel from Alexander Pope's *The Rape of the Lock* (Ariel is also a sprite in Shakespeare's *The Tempest*). The reasoning was presumably that Uranus, as god of the sky and air, would be attended by spirits of the air.

Subsequent names, rather than continuing the airy spirits theme (only Puck and Mab continued the trend), have focused on Herschel's source material. In 1949, the fifth moon, Miranda, was named by its discoverer Gerard Kuiper after a thoroughly mortal character in Shakespeare's *The Tempest*. The current IAU practice is to name moons after characters from Shakespeare's plays and *The Rape of the Lock* (although at present only Ariel, Umbriel, and Belinda have names drawn from the latter; all the rest are from Shakespeare). At first, the outermost moons were all named after characters from one play, *The Tempest*; but with Margaret being named from *Much Ado About Nothing* that trend has ended.

* *The Rape of the Lock* (a poem by Alexander Pope):
  * Ariel, Umbriel, Belinda
* Plays by William Shakespeare:
  * *A Midsummer Night's Dream*: Titania, Oberon, Puck
  * *The Tempest*: (Ariel), Miranda, Caliban, Sycorax, Prospero, Setebos, Stephano, Trinculo, Francisco, Ferdinand
  * *King Lear*: Cordelia
  * *Hamlet*: Ophelia
  * *The Taming of the Shrew*: Bianca
  * *Troilus and Cressida*: Cressida
  * *Othello*: Desdemona
  * *Romeo and Juliet*: Juliet, Mab
  * *The Merchant of Venice*: Portia
  * *As You Like It*: Rosalind
  * *Much Ado About Nothing*: Margaret
  * *The Winter's Tale*: Perdita
  * *Timon of Athens*: Cupid

Some asteroids share names with moons of Uranus: 171 Ophelia, 218 Bianca, 593 Titania, 666 Desdemona, 763 Cupido, and 2758 Cordelia.

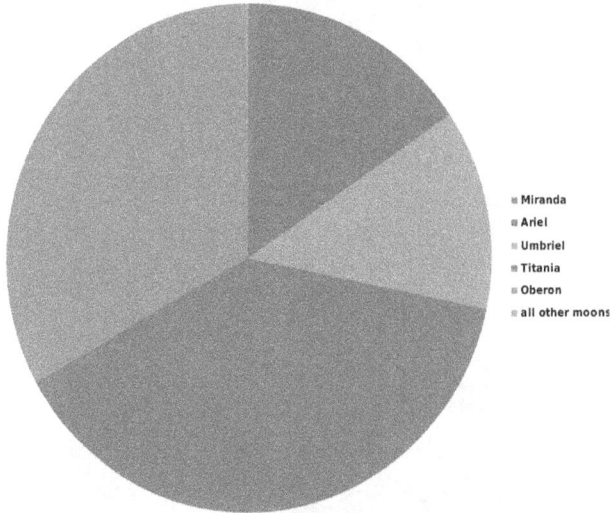

Figure 45: *The relative masses of the Uranian moons. The five rounded moons vary from Miranda at 0.7% to Titania at almost 40% of the total mass. The other moons collectively constitute 0.1%, and are barely visible at this scale.*

Characteristics and groups

The Uranian satellite system is the least massive among those of the giant planets. Indeed, the combined mass of the five major satellites is less than half that of Triton (the seventh-largest moon in the Solar System) alone.[315] The largest of the satellites, Titania, has a radius of 788.9 km, or less than half that of the Moon, but slightly more than that of Rhea, the second-largest moon of Saturn, making Titania the eighth-largest moon in the Solar System. Uranus is about 10,000 times more massive than its moons.[316]

Inner moons

As of 2016, Uranus is known to have 13 inner moons. Their orbits lie inside that of Miranda. All inner moons are intimately connected to the rings of Uranus, which probably resulted from the fragmentation of one or several small inner moons. The two innermost moons (Cordelia and Ophelia) are as shepherds of Uranus's ε ring, whereas the small moon Mab is a source of Uranus's outermost μ ring. There may be two additional small (2–7 km in radius) undiscovered shepherd moons located about 100 km exterior to Uranus' α and β rings.

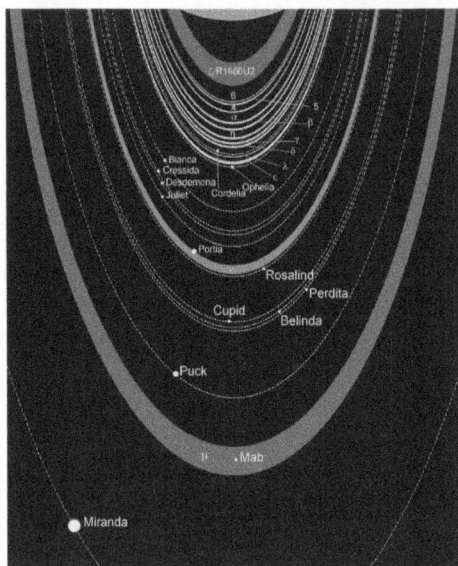

Figure 46: *Schematic of the Uranian moon–ring system*

At 162 km, Puck is the largest of the inner moons of Uranus and the only one imaged by *Voyager 2* in any detail. Puck and Mab are the two outermost inner satellites of Uranus. All inner moons are dark objects; their geometrical albedo is less than 10%. They are composed of water ice contaminated with a dark material—probably radiation-processed organics.

The small inner moons constantly perturb each other. The system is chaotic and apparently unstable. Simulations show that the moons may perturb each other into crossing orbits, which may eventually result in collisions between the moons. Desdemona may collide with either Cressida or Juliet within the next 100 million years.

Large moons

Uranus has five major moons: Miranda, Ariel, Umbriel, Titania, and Oberon. They range in diameter from 472 km for Miranda to 1578 km for Titania. All these moons are relatively dark objects: their geometrical albedo varies between 30 and 50%, whereas their Bond albedo is between 10 and 23%. Umbriel is the darkest moon and Ariel the brightest. The masses of the moons range from 6.7×10^{19} kg (Miranda) to 3.5×10^{21} kg (Titania). For comparison, the Moon has a mass of 7.5×10^{22} kg. The major moons of Uranus are thought to have formed in the accretion disc, which existed around Uranus

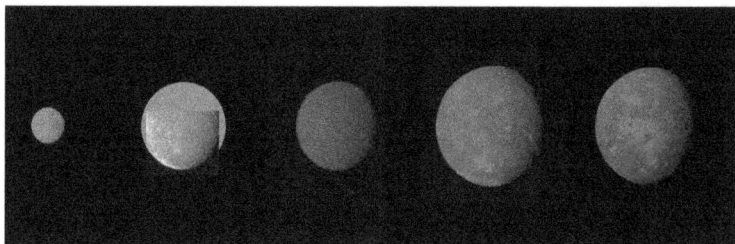

Figure 47: *The five largest moons of Uranus compared at their proper relative sizes and brightnesses. From left to right (in order of increasing distance from Uranus): Miranda, Ariel, Umbriel, Titania, and Oberon.*

Figure 48: *Artist's conception of the Sun's path in the summer sky of a major moon of Uranus (which shares Uranus's axial tilt)*

for some time after its formation or resulted from a large impact suffered by Uranus early in its history.

All major moons comprise approximately equal amounts rock and ice, except Miranda, which is made primarily of ice. The ice component may include ammonia and carbon dioxide. Their surfaces are heavily cratered, though all of them (except Umbriel) show signs of endogenic resurfacing in the form of lineaments (canyons) and, in the case of Miranda, ovoid race-track like structures called coronae. Extensional processes associated with upwelling diapirs are likely responsible for the origin of the coronae. Ariel appears to have the youngest surface with the fewest impact craters, while Umbriel's appears oldest. A past 3:1 orbital resonance between Miranda and Umbriel and a past 4:1 resonance between Ariel and Titania are thought to be responsible for the heating that caused substantial endogenic activity on Miranda and Ariel. One

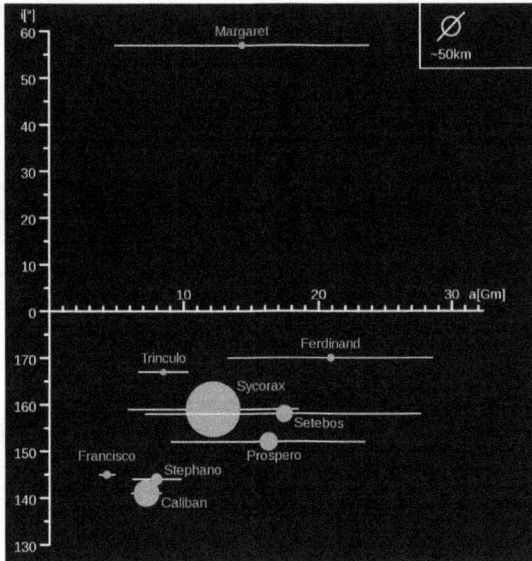

Figure 49: *Irregular moons of Uranus. The X axis is labeled in Gm (million km) and in the fraction of the Hill sphere's radius. The eccentricity is represented by the yellow segments (extending from the pericentre to the apocentre) with the inclination represented on the Y axis.*

piece of evidence for such a past resonance is Miranda's unusually high orbital inclination (4.34°) for a body so close to the planet. The largest Uranian moons may be internally differentiated, with rocky cores at their centers surrounded by ice mantles. Titania and Oberon may harbor liquid water oceans at the core/mantle boundary. The major moons of Uranus are airless bodies. For instance, Titania was shown to possess no atmosphere at a pressure larger than 10–20 nanobar.

The path of the Sun in the local sky over the course of a local day during Uranus's and its major moons' summer solstice is quite different from that seen on most other Solar System worlds. The major moons have almost exactly the same rotational axial tilt as Uranus (their axes are parallel to that of Uranus). The Sun would appear to follow a circular path around Uranus's celestial pole in the sky, at the closest about 7 degrees from it.[317] Near the equator, it would be seen nearly due north or due south (depending on the season). At latitudes higher than 7°, the Sun would trace a circular path about 15 degrees in diameter in the sky, and never set.

Irregular moons

As of 2005 Uranus is known to have nine irregular moons, which orbit it at a distance much greater than that of Oberon, the furthest of the large moons. All the irregular moons are probably captured objects that were trapped by Uranus soon after its formation. The diagram illustrates the orbits of those irregular moons discovered so far. The moons above the X axis are prograde, those beneath are retrograde. The radius of the Uranian Hill sphere is approximately 73 million km.

Uranus's irregular moons range in size from 120–200 km (Sycorax) to about 20 km (Trinculo). Unlike Jupiter's irregulars, Uranus's show no correlation of axis with inclination. Instead, the retrograde moons can be divided into two groups based on axis/orbital eccentricity. The inner group includes those satellites closer to Uranus (a < 0.15 r_H) and moderately eccentric ($\sim$0.2), namely Francisco, Caliban, Stephano, and Trinculo. The outer group (a > 0.15 r_H) includes satellites with high eccentricity ($\sim$0.5): Sycorax, Prospero, Setebos, and Ferdinand.

The intermediate inclinations 60° < i < 140° are devoid of known moons due to the Kozai instability. In this instability region, solar perturbations at apoapse cause the moons to acquire large eccentricities that lead to collisions with inner satellites or ejection. The lifetime of moons in the instability region is from 10 million to a billion years.

Margaret is the only known irregular prograde moon of Uranus, and it currently has the most eccentric orbit of any moon in the Solar System, though Neptune's moon Nereid has a higher mean eccentricity. As of 2008, Margaret's eccentricity is 0.7979.

List

Key

Inner moons	† Major moons	‡ Irregular moons (retrograde)	° Irregular moon (prograde)

The Uranian moons are listed here by orbital period, from shortest to longest. Moons massive enough for their surfaces to have collapsed into a spheroid are highlighted in light blue and bolded. Irregular moons with retrograde orbits are shown in dark grey. Margaret, the only known irregular moon of Uranus with a prograde orbit, is shown in light grey.

Uranian moons

Order[318]	Label[319]	Name	Pronunciation (key)	Image	Diameter (km)[320]	Mass (×10^18 kg)[321]	Semi-major axis (km)	Orbital period (d)[322]	Inclination (°)[323]	Eccentricity	Discovery year	Discoverer
1	VI	Cordelia	/kɔːrˈdiːliə/		40 ± 6 (50 × 36)	0.044	49770	0.335034	0.08479°	0.00026	1986	Terrile (Voyager 2)
2	VII	Ophelia	/oʊˈfiːliə/		43 ± 8 (54 × 38)	0.053	53790	0.376400	0.1036°	0.00992	1986	Terrile (Voyager 2)
3	VIII	Bianca	/biˈɑːŋkə/		51 ± 4 (64 × 46)	0.092	59170	0.434579	0.193°	0.00092	1986	Smith (Voyager 2)
4	IX	Cressida	/ˈkrɛsɪdə/		80 ± 4 (92 × 74)	0.34	61780	0.463570	0.006°	0.00036	1986	Synnott (Voyager 2)
5	X	Desdemona	/ˌdɛzdɪˈmoʊnə/		64 ± 8 (90 × 54)	0.18	62680	0.473650	0.11125°	0.00013	1986	Synnott (Voyager 2)
6	XI	Juliet	/ˈdʒuːliet/		94 ± 8 (150 × 74)	0.56	64350	0.493065	0.065°	0.00066	1986	Synnott (Voyager 2)
7	XII	Portia	/ˈpɔːrʃə/		135 ± 8 (156 × 126)	1.70	66090	0.513196	0.059°	0.00005	1986	Synnott (Voyager 2)
8	XIII	Rosalind	/ˈrɒzəlɪnd/		72 ± 12	0.25	69940	0.558460	0.279°	0.00011	1986	Synnott (Voyager 2)
9	XXVII	Cupid	/ˈkjuːpɪd/		≈ 18	0.0038	74800	0.618	0.1°	0.0013	2003	Showalter and Lissauer

#	Desig.	Name	Pronunciation	Image	Size (km)		Distance (km)	Period (days)	Inclination	Eccentricity	Year	Discoverer
10	XIV	Belinda	/bɪˈlɪndə/		90 ± 16 (128 × 64)	0.49	75260	0.623527	0.031°	0.00007	1986	Synnott (Voyager 2)
11	XXV	Perdita	/pɜˈdiːtə/		30 ± 6	0.018	76400	0.638	0.0°	0.0012	1999	Karkoschka (Voyager 2)
12	XV	Puck	/ˈpʌk/		162 ± 4	2.90	86010	0.761833	0.3192°	0.00012	1985	Synnott (Voyager 2)
13	XXVI	Mab	/ˈmæb/		≈ 25	0.01	97700	0.923	0.1335°	0.0025	2003	Showalter and Lissauer
14	V	†Miranda	/mɪˈrændə/		471.6 ± 1.4 (481 × 468 × 466)	65.9±7.5	129390	1.413479	4.232°	0.0013	1948	Kuiper
15	I	†Ariel	/ˈɛəriəl/		1157.8±1.2 (1162 × 1156 × 1155)	1353±120	191020	2.520379	0.260°	0.0012	1851	Lassell
16	II	†Umbriel	/ˈʌmbriəl/		1169.4±5.6	1172±135	266300	4.144177	0.205°	0.0039	1851	Lassell
17	III	†Titania	/tɪˈteɪniə/		1576.8±1.2	3527±90	435910	8.705872	0.340°	0.0011	1787	Herschel

					1522.8±5.2	3014±75	583520	13.463239	0.058°	0.0014	1787	Herschel
18	IV	†Oberon	/ˈoʊbərɒn/									
19	XXII	‡Francisco	/frænˈsɪskoʊ/		≈22	0.0072	4276000	−266.56	147.459°	0.1459	2003[324]	Holman et al.
20	XVI	‡Caliban	/ˈkælɪbæn/		≈72	0.25	7230000	−579.50	139.885°	0.1587	1997	Gladman et al.
21	XX	‡Stephano	/ˈstɛfənoʊ/		≈32	0.022	8002000	−676.50	141.873°	0.2292	1999	Gladman et al.
22	XXI	‡Trinculo	/ˈtrɪŋkjʊloʊ/		≈18	0.0039	8571000	−758.10	166.252°	0.2200	2001	Holman et al.
23	XVII	‡Sycorax	/ˈsɪkəræks/		165+36 -42	2.30	12179000	−1283.4	152.456°	0.5224	1997	Nicholson et al.
24	XXIII	°Margaret	/ˈmɑːrɡərɪt/		≈20	0.0054	14345000	1694.8	51.455°	0.6608	2003	Sheppard and Jewitt
25	XVIII	‡Prospero	/ˈprɒspəroʊ/		≈50	0.085	16418000	−1992.8	146.017°	0.4448	1999	Holman et al.
26	XIX	‡Setebos	/ˈsɛtɪbɒs/		≈48	0.075	17459000	−2202.3	145.883°	0.5914	1999	Kavelaars et al.
27	XXIV	‡Ferdinand	/ˈfɜːrdmænd/		≈20	0.0054	20900000	−2823.4	167.346°	0.3682	2003	Holman et al.

Sources: NASA/NSSDC, Sheppard, et al. 2005. For the recently discovered outer irregular moons (Francisco through Ferdinand) the most accurate orbital data can be generated with the Natural Satellites Ephemeris Service. The irregulars are significantly perturbed by the Sun.

External links

Wikimedia Commons has media related to *Moons of Uranus.*

- Simulation Showing the location of Uranus's Moons[325]
- "Uranus: Moons"[326]. NASA's Solar System Exploration. Retrieved 20 December 2008.
- "NASA's Hubble Discovers New Rings and Moons Around Uranus"[327]. Space Telescope Science Institute. 22 December 2005. Retrieved 20 December 2008.
- Sheppard, Scott. "Uranus's Known Satellites"[328]. Retrieved 20 December 2008.
- Gazetteer of Planetary Nomenclature—Uranus (USGS)[329]

<indicator name="featured-star"> ⭐ </indicator>

Planetary rings

Rings of Uranus

The **rings of Uranus** are a system of rings around the planet Uranus, intermediate in complexity between the more extensive set around Saturn and the simpler systems around Jupiter and Neptune. The rings of Uranus were discovered on March 10, 1977, by James L. Elliot, Edward W. Dunham, and Jessica Mink. William Herschel had also reported observing rings in 1789; modern astronomers are divided on whether he could have seen them, as they are very dark and faint.[330]

By 1978, nine distinct rings were identified. Two additional rings were discovered in 1986 in images taken by the *Voyager 2* spacecraft, and two outer rings were found in 2003–2005 in Hubble Space Telescope photos. In the order of increasing distance from the planet the 13 known rings are designated 1986U2R/ζ, 6, 5, 4, α, β, η, γ, δ, λ, ε, ν and μ. Their radii range from about 38,000 km for the 1986U2R/ζ ring to about 98,000 km for the μ ring. Additional faint dust bands and incomplete arcs may exist between the main rings. The rings are extremely dark—the Bond albedo of the rings' particles does not exceed 2%. They are probably composed of water ice with the addition of some dark radiation-processed organics.

The majority of Uranus's rings are opaque and only a few kilometers wide. The ring system contains little dust overall; it consists mostly of large bodies 0.2–20 m in diameter. Some rings are optically thin: the broad and faint 1986U2R/ζ, μ and ν rings are made of small dust particles, while the narrow and faint λ ring also contains larger bodies. The relative lack of dust in the ring system may be due to aerodynamic drag from the extended Uranian exosphere.

The rings of Uranus are thought to be relatively young, and not more than 600 million years old. The Uranian ring system probably originated from the collisional fragmentation of several moons that once existed around the planet.

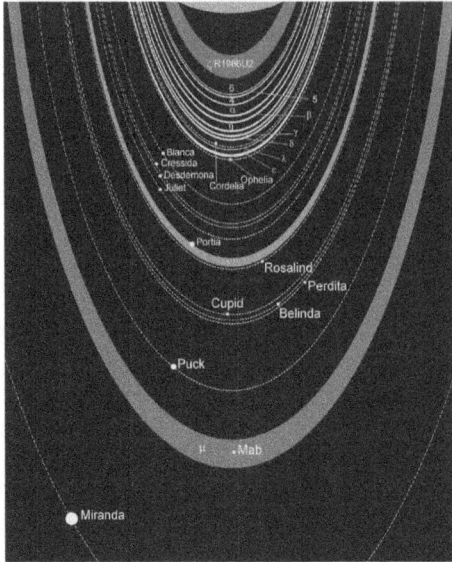

Figure 50: *The scheme of Uranus's ring-moon system. Solid lines denote rings; dashed lines denote orbits of moons.*

After colliding, the moons probably broke up into many particles, which survived as narrow and optically dense rings only in strictly confined zones of maximum stability.

The mechanism that confines the narrow rings is not well understood. Initially it was assumed that every narrow ring had a pair of nearby shepherd moons corralling them into shape. In 1986 *Voyager 2* discovered only one such shepherd pair (Cordelia and Ophelia) around the brightest ring (ε).

Discovery

The first mention of a Uranian ring system comes from William Herschel's notes detailing his observations of Uranus in the 18th century, which include the following passage: "February 22, 1789: A ring was suspected". Herschel drew a small diagram of the ring and noted that it was "a little inclined to the red". The Keck Telescope in Hawaii has since confirmed this to be the case, at least for the ν ring. Herschel's notes were published in a Royal Society journal in 1797. In the two centuries between 1797 and 1977 the rings are rarely mentioned, if at all. This casts serious doubt on whether Herschel could have seen anything of the sort while hundreds of other astronomers saw nothing. It has been claimed that Herschel gave accurate descriptions of the ε ring's size relative to Uranus, its changes as Uranus travelled around the Sun, and its color.

The definitive discovery of the Uranian rings was made by astronomers James L. Elliot, Edward W. Dunham, and Jessica Mink on March 10, 1977, using the Kuiper Airborne Observatory, and was serendipitous. They planned to use the occultation of the star SAO 158687 by Uranus to study the planet's atmosphere. When their observations were analyzed, they found that the star disappeared briefly from view five times both before and after it was eclipsed by the planet. They deduced that a system of narrow rings was present. The five occultation events they observed were denoted by the Greek letters α, β, γ, δ and ε in their papers. These designations have been used as the rings' names since then. Later they found four additional rings: one between the β and γ rings and three inside the α ring. The former was named the η ring. The latter were dubbed rings 4, 5 and 6—according to the numbering of the occultation events in one paper. Uranus's ring system was the second to be discovered in the Solar System, after that of Saturn.

The rings were directly imaged when the *Voyager 2* spacecraft flew through the Uranian system in 1986. Two more faint rings were revealed, bringing the total to eleven. The Hubble Space Telescope detected an additional pair of previously unseen rings in 2003–2005, bringing the total number known to 13. The discovery of these outer rings doubled the known radius of the ring system. Hubble also imaged two small satellites for the first time, one of which, Mab, shares its orbit with the outermost newly discovered μ ring.

General properties

As currently understood, the ring system of Uranus comprises thirteen distinct rings. In order of increasing distance from the planet they are: 1986U2R/ζ, 6, 5, 4, α, β, η, γ, δ, λ, ε, ν, μ rings. They can be divided into three groups: nine narrow main rings (6, 5, 4, α, β, η, γ, δ, ε), two dusty rings (1986U2R/ζ, λ) and two outer rings (ν, μ). The rings of Uranus consist mainly of macroscopic particles and little dust, although dust is known to be present in 1986U2R/ζ, η, δ, λ, ν and μ rings. In addition to these well-known rings, there may be numerous optically thin dust bands and faint rings between them. These faint rings and dust bands may exist only temporarily or consist of a number of separate arcs, which are sometimes detected during occultations. Some of them became visible during a series of ring plane-crossing events in 2007. A number of dust bands between the rings were observed in forward-scattering[331] geometry by *Voyager 2*. All rings of Uranus show azimuthal brightness variations.

The rings are made of an extremely dark material. The geometric albedo of the ring particles does not exceed 5–6%, while the Bond albedo is

Figure 51: *Uranus's inner rings. The bright outer
ring is the epsilon ring; eight other rings are visible*

even lower—about 2%. The rings particles demonstrate a steep opposition
surge—an increase of the albedo when the phase angle is close to zero. This
means that their albedo is much lower when they are observed slightly off the
opposition.[332] The rings are slightly red in the ultraviolet and visible parts of
the spectrum and grey in near-infrared. They exhibit no identifiable spectral
features. The chemical composition of the ring particles is not known. They
cannot be made of pure water ice like the rings of Saturn because they are
too dark, darker than the inner moons of Uranus. This indicates that they are
probably composed of a mixture of the ice and a dark material. The nature of
this material is not clear, but it may be organic compounds considerably darkened by the charged particle irradiation from the Uranian magnetosphere. The
rings' particles may consist of a heavily processed material which was initially
similar to that of the inner moons.

As a whole, the ring system of Uranus is unlike either the faint dusty rings of
Jupiter or the broad and complex rings of Saturn, some of which are composed
of very bright material—water ice. There are similarities with some parts of the
latter ring system; the Saturnian F ring and the Uranian ε ring are both narrow,
relatively dark and are shepherded by a pair of moons. The newly discovered
outer ν and μ rings of Uranus are similar to the outer G and E rings of Saturn.
Narrow ringlets existing in the broad Saturnian rings also resemble the narrow

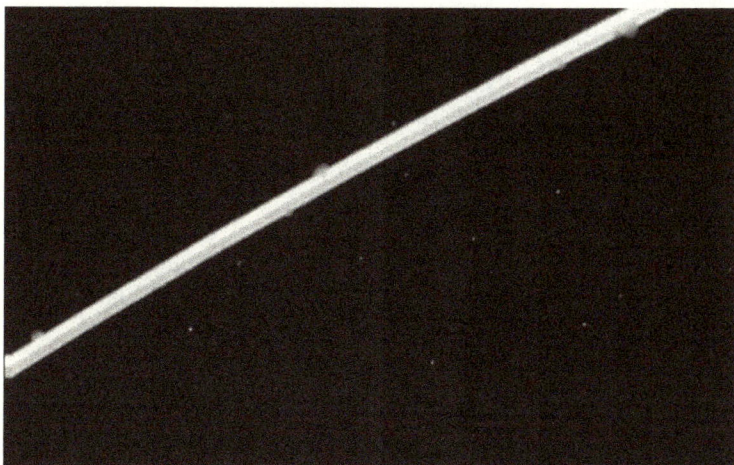

Figure 52: *A close-up view of the ε ring of Uranus*

rings of Uranus. In addition, dust bands observed between the main rings of
Uranus may be similar to the rings of Jupiter. In contrast, the Neptunian ring
system is quite similar to that of Uranus, although it is less complex, darker
and contains more dust; the Neptunian rings are also positioned further from
the planet.

Narrow main rings

ε ring

The ε ring is the brightest and densest part of the Uranian ring system, and
is responsible for about two-thirds of the light reflected by the rings. While
it is the most eccentric of the Uranian rings, it has negligible orbital inclination. The ring's eccentricity causes its brightness to vary over the course of its
orbit. The radially integrated brightness of the ε ring is highest near apoapsis
and lowest near periapsis. The maximum/minimum brightness ratio is about
2.5–3.0. These variations are connected with the variations of the ring width,
which is 19.7 km at the periapsis and 96.4 km at the apoapsis. As the ring becomes wider, the amount of shadowing between particles decreases and more
of them come into view, leading to higher integrated brightness. The width
variations were measured directly from *Voyager 2* images, as the ε ring was
one of only two rings resolved by Voyager's cameras. Such behavior indicates
that the ring is not optically thin. Indeed, occultation observations conducted
from the ground and the spacecraft showed that its normal optical depth[333]

Figure 53: *A close-up view of the (from top to bottom) δ, γ, η, β and α rings of Uranus. The resolved η ring demonstrates the optically thin broad component.*

varies between 0.5 and 2.5, being highest near the periapsis. The equivalent depth[334] of the ε ring is around 47 km and is invariant around the orbit.

The geometric thickness of the ε ring is not precisely known, although the ring is certainly very thin—by some estimates as thin as 150 m. Despite such infinitesimal thickness, it consists of several layers of particles. The ε ring is a rather crowded place with a filling factor near the apoapsis estimated by different sources at from 0.008 to 0.06. The mean size of the ring particles is 0.2–20.0 m, and the mean separation is around 4.5 times their radius. The ring is almost devoid of dust, possibly due to the aerodynamic drag from Uranus's extended atmospheric corona. Due to its razor-thin nature the ε ring is invisible when viewed edge-on. This happened in 2007 when a ring plane-crossing was observed.

The *Voyager 2* spacecraft observed a strange signal from the ε ring during the radio occultation experiment. The signal looked like a strong enhancement of the forward-scattering at the wavelength 3.6 cm near ring's apoapsis. Such strong scattering requires the existence of a coherent structure. That the ε ring does have such a fine structure has been confirmed by many occultation observations. The ε ring seems to consist of a number of narrow and optically dense ringlets, some of which may have incomplete arcs.

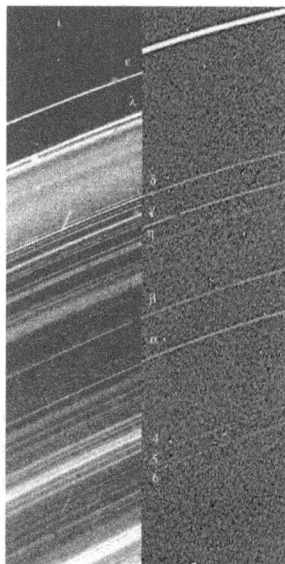

Figure 54: *Comparison of the Uranian rings in forward-scattered and back-scattered light (images obtained by Voyager 2 in 1986)*

The ε ring is known to have interior and exterior shepherd moons—Cordelia and Ophelia, respectively. The inner edge of the ring is in 24:25 resonance with Cordelia, and the outer edge is in 14:13 resonance with Ophelia. The masses of the moons need to be at least three times the mass of the ring to confine it effectively. The mass of the ε ring is estimated to be about 10^{16} kg.

δ ring

The δ ring is circular and slightly inclined. It shows significant unexplained azimuthal variations in normal optical depth and width. One possible expla-nation is that the ring has an azimuthal wave-like structure, excited by a small moonlet just inside it. The sharp outer edge of the δ ring is in 23:22 reso-nance with Cordelia. The δ ring consists of two components: a narrow opti-cally dense component and a broad inward shoulder with low optical depth. The width of the narrow component is 4.1–6.1 km and the equivalent depth is about 2.2 km, which corresponds to a normal optical depth of about 0.3–0.6. The ring's broad component is about 10–12 km wide and its equivalent depth is close to 0.3 km, indicating a low normal optical depth of 3×10^{-2}. This is known only from occultation data because *Voyager 2's* imaging experiment failed to resolve the δ ring. When observed in forward-scattering geometry by *Voyager 2*, the δ ring appeared relatively bright, which is compatible with the presence of dust in its broad component. The broad component is geometrically thicker than the narrow component. This is supported by the observations of a ring plane-crossing event in 2007, when the δ ring remained visible, which is consistent with the behavior of a simultaneously geometrically thick and optically thin ring.

γ ring

The γ ring is narrow, optically dense and slightly eccentric. Its orbital inclination is almost zero. The width of the ring varies in the range 3.6–4.7 km, although equivalent optical depth is constant at 3.3 km. The normal optical depth of the γ ring is 0.7–0.9. During a ring plane-crossing event in 2007 the γ ring disappeared, which means it is geometrically thin like the ε ring and devoid of dust. The width and normal optical depth of the γ ring show significant azimuthal variations. The mechanism of confinement of such a narrow ring is not known, but it has been noticed that the sharp inner edge of the γ ring is in a 6:5 resonance with Ophelia.

η ring

The η ring has zero orbital eccentricity and inclination. Like the δ ring, it consists of two components: a narrow optically dense component and a broad outward shoulder with low optical depth. The width of the narrow component is 1.9–2.7 km and the equivalent depth is about 0.42 km, which corresponds to the normal optical depth of about 0.16–0.25. The broad component is about 40 km wide and its equivalent depth is close to 0.85 km, indicating a low normal optical depth of 2×10^{-2}. It was resolved in *Voyager 2* images. In forward-scattered light, the η ring looked bright, which indicated the presence of a considerable amount of dust in this ring, probably in the broad component. The broad component is much thicker (geometrically) than the narrow one. This conclusion is supported by the observations of a ring plane-crossing event in 2007, when the η ring demonstrated increased brightness, becoming the second brightest feature in the ring system. This is consistent with the behavior of a geometrically thick but simultaneously optically thin ring. Like the majority of other rings, the η ring shows significant azimuthal variations in the normal optical depth and width. The narrow component even vanishes in some places.

α and β rings

After the ε ring, the α and β rings are the brightest of Uranus's rings. Like the ε ring, they exhibit regular variations in brightness and width. They are brightest and widest 30° from the apoapsis and dimmest and narrowest 30° from the periapsis. The α and β rings have sizable orbital eccentricity and non-negligible inclination. The widths of these rings are 4.8–10 km and 6.1–11.4 km, respectively. The equivalent optical depths are 3.29 km and 2.14 km, resulting in normal optical depths of 0.3–0.7 and 0.2–0.35, respectively. During a ring plane-crossing event in 2007 the rings disappeared, which means they are geometrically thin like the ε ring and devoid of dust. The same event revealed a thick and optically thin dust band just outside the β ring, which was also observed earlier by *Voyager 2*. The masses of the α and β rings are estimated to be about 5×10^{15} kg (each)—half the mass of the ε ring.

Rings 6, 5 and 4

Rings 6, 5 and 4 are the innermost and dimmest of Uranus's narrow rings. They are the most inclined rings, and their orbital eccentricities are the largest excluding the ε ring. In fact, their inclinations (0.06°, 0.05° and 0.03°) were large enough for *Voyager 2* to observe their elevations above the Uranian equatorial plane, which were 24–46 km. Rings 6, 5 and 4 are also the narrowest rings of Uranus, measuring 1.6–2.2 km, 1.9–4.9 km and 2.4–4.4 km wide, respectively. Their equivalent depths are 0.41 km, 0.91 and 0.71 km resulting in normal optical depth 0.18–0.25, 0.18–0.48 and 0.16–0.3. They were not visible during a ring plane-crossing event in 2007 due to their narrowness and lack of dust.

Dusty rings

λ ring

The λ ring was one of two rings discovered by *Voyager 2* in 1986. It is a narrow, faint ring located just inside the ε ring, between it and the shepherd moon Cordelia. This moon clears a dark lane just inside the λ ring. When viewed in back-scattered light,[335] the λ ring is extremely narrow—about 1–2 km—and has the equivalent optical depth 0.1–0.2 km at the wavelength 2.2 μm. The normal optical depth is 0.1–0.2. The optical depth of the λ ring shows strong wavelength dependence, which is atypical for the Uranian ring system. The equivalent depth is as high as 0.36 km in the ultraviolet part of the spectrum, which explains why λ ring was initially detected only in UV stellar occultations by *Voyager 2*. The detection during a stellar occultation at the wavelength 2.2 μm was only announced in 1996.

The appearance of the λ ring changed dramatically when it was observed in forward-scattered light in 1986. In this geometry the ring became the brightest feature of the Uranian ring system, outshining the ε ring. This observation, together with the wavelength dependence of the optical depth, indicates that the λ ring contains significant amount of micrometer-sized dust. The normal

Figure 55: *A long-exposure, high phase angle (172.5°) Voyager 2 image of Uranus's inner rings. In forward-scattered light, dust bands not visible in other images can be seen, as well as the recognized rings.*

optical depth of this dust is 10^{-4}–10^{-3}. Observations in 2007 by the Keck telescope during the ring plane-crossing event confirmed this conclusion, because the λ ring became one of the brightest features in the Uranian ring system.

Detailed analysis of the *Voyager 2* images revealed azimuthal variations in the brightness of the λ ring. The variations appear to be periodic, resembling a standing wave. The origin of this fine structure in the λ ring remains a mystery.

1986U2R/ζ ring

In 1986 *Voyager 2* detected a broad and faint sheet of material inward of ring 6. This ring was given the temporary designation 1986U2R. It had a normal optical depth of 10^{-3} or less and was extremely faint. It was visible only in a single *Voyager 2* image. The ring was located between 37,000 and 39,500 km from the centre of Uranus, or only about 12,000 km above the clouds. It was not observed again until 2003–2004, when the Keck telescope found a broad and faint sheet of material just inside ring 6. This ring was dubbed the ζ ring. The position of the recovered ζ ring differs significantly from that observed in 1986. Now it is situated between 37,850 and 41,350 km from the centre of the planet. There is an inward gradually fading extension reaching to at

Figure 56: *The discovery image of the 1986U2R ring*

least 32,600 km, or possibly even to 27,000 km—to the atmosphere of Uranus. These extensions are labelled as the ζ_c and ζ_{cc} rings respectively.

The ζ ring was observed again during the ring plane-crossing event in 2007 when it became the brightest feature of the ring system, outshining all other rings combined. The equivalent optical depth of this ring is near 1 km (0.6 km for the inward extension), while the normal optical depth is again less than 10^{-3}. Rather different appearances of the 1986U2R and ζ rings may be caused by different viewing geometries: back-scattering geometry in 2003–2007 and side-scattering geometry in 1986. Changes during the past 20 years in the distribution of dust, which is thought to predominate in the ring, cannot be ruled out.

Other dust bands

In addition to the 1986U2R/ζ and λ rings, there are other extremely faint dust bands in the Uranian ring system. They are invisible during occultations because they have negligible optical depth, though they are bright in forward-scattered light. *Voyager 2*'s images of forward-scattered light revealed the existence of bright dust bands between the λ and δ rings, between the η and β rings, and between the α ring and ring 4. Many of these bands were detected again in 2003–2004 by the Keck Telescope and during the 2007 ring-plane crossing event in backscattered light, but their precise locations and relative

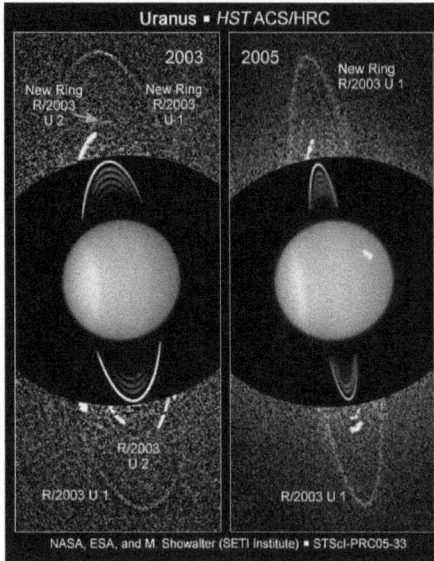

Figure 57: *The μ and ν rings of Uranus (R/2003 U1 and U2)
in Hubble Space Telescope images from 2005*

brightnesses were different from during the *Voyager* observations. The normal optical depth of the dust bands is about 10^{-5} or less. The dust particle size distribution is thought to obey a power law with the index $p = 2.5 \pm 0.5$.

In addition to separate dust bands the system of Uranian rings appears to be immersed into wide and faint sheet of dust with the normal optical depth not exceeding 10^{-3}.

Outer ring system

In 2003–2005, the Hubble Space Telescope detected a pair of previously unknown rings, now called the outer ring system, which brought the number of known Uranian rings to 13. These rings were subsequently named the μ and ν rings. The μ ring is the outermost of the pair, and is twice the distance from the planet as the bright η ring. The outer rings differ from the inner narrow rings in a number of respects. They are broad, 17,000 and 3,800 km wide, respectively, and very faint. Their peak normal optical depths are 8.5×10^{-6} and 5.4×10^{-6}, respectively. The resulting equivalent optical depths are 0.14 km and 0.012 km. The rings have triangular radial brightness profiles.

The peak brightness of the μ ring lies almost exactly on the orbit of the small Uranian moon Mab, which is probably the source of the ring's particles. The

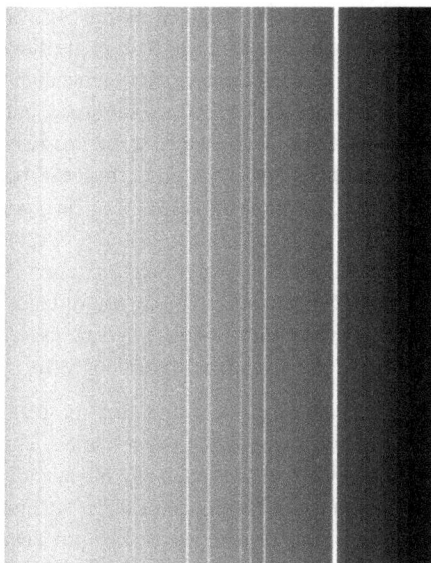

Figure 58: *An enhanced-color schematic of*
the inner rings derived from Voyager 2 images

ν ring is positioned between Portia and Rosalind and does not contain any moons inside it. A reanalysis of the *Voyager 2* images of forward-scattered light clearly reveals the μ and ν rings. In this geometry the rings are much brighter, which indicates that they contain much micrometer-sized dust. The outer rings of Uranus may be similar to the G and E rings of Saturn as E ring is extremely broad and receives dust from Enceladus.

The μ ring may consist entirely of dust, without any large particles at all. This hypothesis is supported by observations performed by the Keck telescope, which failed to detect the μ ring in the near infrared at 2.2 μm, but detected the ν ring. This failure means that the μ ring is blue in color, which in turn indicates that very small (submicrometer) dust predominates within it. The dust may be made of water ice. In contrast, the ν ring is slightly red in color.

Dynamics and origin

An outstanding problem concerning the physics governing the narrow Uranian rings is their confinement. Without some mechanism to hold their particles together, the rings would quickly spread out radially. The lifetime of the Uranian rings without such a mechanism cannot be more than 1 million years. The most widely cited model for such confinement, proposed initially by Goldreich and Tremaine, is that a pair of nearby moons, outer and inner shepherds, interact gravitationally with a ring and act like sinks and donors, respectively, for excessive and insufficient angular momentum (or equivalently, energy). The shepherds thus keep ring particles in place, but gradually move away from the ring themselves. To be effective, the masses of the shepherds should exceed the mass of the ring by at least a factor of two to three. This mechanism is known to be at work in the case of the ε ring, where Cordelia and Ophelia serve as shepherds. Cordelia is also the outer shepherd of the δ ring, and Ophelia is the outer shepherd of the γ ring. No moon larger than 10 km is known in the vicinity of other rings. The current distance of Cordelia and Ophelia from the ε ring can be used to estimate the ring's age. The calculations show that the ε ring cannot be older than 600 million years.

Since the rings of Uranus appear to be young, they must be continuously renewed by the collisional fragmentation of larger bodies. The estimates show that the lifetime against collisional disruption of a moon with the size like that of Puck is a few billion years. The lifetime of a smaller satellite is much shorter. Therefore, all current inner moons and rings can be products of disruption of several Puck-sized satellites during the last four and half billion years. Every such disruption would have started a collisional cascade that quickly ground almost all large bodies into much smaller particles, including dust. Eventually the majority of mass was lost, and particles survived only in positions that were stabilized by mutual resonances and shepherding. The end product of such a disruptive evolution would be a system of narrow rings. A few moonlets must still be embedded within the rings at present. The maximum size of such moonlets is probably around 10 km.

The origin of the dust bands is less problematic. The dust has a very short lifetime, 100–1000 years, and should be continuously replenished by collisions between larger ring particles, moonlets and meteoroids from outside the Uranian system. The belts of the parent moonlets and particles are themselves invisible due to their low optical depth, while the dust reveals itself in forward-scattered light. The narrow main rings and the moonlet belts that create dust bands are expected to differ in particle size distribution. The main rings have more centimeter to meter-sized bodies. Such a distribution increases the surface area of the material in the rings, leading to high optical density in backscattered light. In contrast, the dust bands have relatively few large particles, which results in low optical depth.

Exploration

The rings were thoroughly investigated by the Voyager 2 spacecraft in January 1986. Two new faint rings—λ and 1986U2R—were discovered bringing the total number then known to eleven. Rings were studied by analyzing results of radio, ultraviolet and optical occultations. *Voyager 2* observed the rings in different geometries relative to the sun, producing images with back-scattered, forward-scattered and side-scattered light. Analysis of these images allowed derivation of the complete phase function, geometrical and Bond albedo of ring particles. Two rings—ε and η—were resolved in the images revealing a complicated fine structure. Analysis of Voyager's images also led to discovery of eleven inner moons of Uranus, including the two shepherd moons of the ε ring—Cordelia and Ophelia.

List of properties

This table summarizes the properties of the planetary ring system of Uranus.

Ring name	Radius (km)[336]	Width (km)	Eq. depth (km)[337]	N. Opt. depth[338]	Thickness (m)[339]	Ecc.[340] </-ref>	Incl. (°)	Notes
ζcc	26 840–34 890	8 000	0.8	∼ 0.001	?	?	?	Inward extension of the ζc ring
ζc	34 890–37 850	3 000	0.6	∼ 0.01	?	?	?	Inward extension of the ζ ring
1986U2R	37 000–39 500	2 500	<2.5	< 0.01	?	?	?	Faint dusty ring
ζ	37 850–41 350	3 500	1	∼ 0.01	?	?	?	
6	41 837	1.6–2.2	0.41	0.18–0.25	?	0.01	0.062	
5	42 234	1.9–4.9	0.91	0.18–0.48	?	0.019	0.054	
4	42 570	2.4–4.4	0.71	0.16–0.30	?	0.011	0.032	
α	44 718	4.8–10.0	3.39	0.3–0.7	?	0.008	0.015	
β	45 661	6.1–11.4	2.14	0.20–0.35	?	0.004	0.005	
η	47 175	1.9–2.7	0.42	0.16–0.25	?	0	0.001	
ηc	47 176	40	0.85	0.2	?	0	0.001	Outward broad component of the η ring
γ	47 627	3.6–4.7	3.3	0.7–0.9	150?	0.001	0.002	
δc	48 300	10–12	0.3	0.3	?	0	0.001	Inward broad component of the δ ring
δ	48 300	4.1–6.1	2.2	0.3–0.6	?	0	0.001	
λ	50 023	1–2	0.2	0.1–0.2	?	0?	0?	Faint dusty ring
ε	51 149	19.7–96.4	47	0.5–2.5	150?	0.079	0	Shepherded by Cordelia and Ophelia
ν	66 100–69 900	3 800	0.012	0.000054	?	?	?	Between Portia and Rosalind, peak brightness at 67 300 km
μ	86 000–103 000	17 000	0.14	0.000085	?	?	?	At Mab, peak brightness at 97 700 km

External links

- Uranus' Rings[341] by NASA's Solar System Exploration[342]
- Uranus Rings Fact Sheet[343]
- Hubble Discovers Giant Rings and New Moons Encircling Uranus[344] – Hubble Space Telescope news release (22 December 2005)
- Gazetteer of Planetary Nomenclature – Ring and Ring Gap Nomenclature (Uranus), USGS[345]

<indicator name="featured-star"> ⭐ </indicator>

Exploration

Exploration of Uranus

The **exploration of Uranus** has, to date, been solely through telescopes and NASA's *Voyager 2* spacecraft, which made its closest approach to Uranus on January 24, 1986. *Voyager 2* discovered 10 moons, studied the planet's cold atmosphere, and examined its ring system, discovering two new rings. It also imaged Uranus' five large moons, revealing that their surfaces are covered with impact craters and canyons.

A number of dedicated exploratory missions to Uranus have been proposed, but as of 2017[346] none have been approved.

Voyager 2

Voyager 2 made its closest approach to Uranus on January 24, 1986, coming within 81,500 km (50,600 miles) of the planet's cloud tops. This was the probe's first solo planetary flyby, since *Voyager 1* ended its tour of the outer planets at Saturn's moon Titan.

Uranus is the third-largest planet in the Solar System. It orbits the Sun at a distance of about 2.8 billion kilometers (1.7 billion miles) and completes one orbit every 84 years. The length of a day on Uranus as measured by *Voyager 2* is 17 hours and 14 minutes. Uranus is distinguished by the fact that it is tipped on its side. Its unusual position is thought to be the result of a collision with a planet-sized body early in the Solar System's history. Given its odd orientation, with its polar regions exposed to sunlight or darkness for long periods and *Voyager 2* set to arrive around the time of Uranus's solstice, scientists were not sure what to expect at Uranus.

The presence of a magnetic field at Uranus was not known until *Voyager 2*'s arrival. The intensity of the field is roughly comparable to that of Earth's, though it varies much more from point to point because of its large offset from

Figure 59: *A colour photograph of Uranus, taken by Voyager 2 in 1986 as it headed towards the planet Neptune*

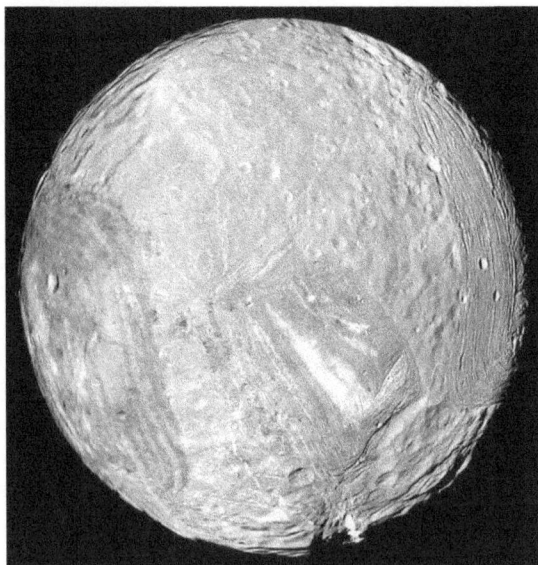

Figure 60: *The Uranian moon Miranda, imaged by Voyager 2*

Figure 61: *A Voyager 2 image of the Uranian dark rings*

the center of Uranus. The peculiar orientation of the magnetic field suggests that the field is generated at an intermediate depth in the interior where the pressure is high enough for water to become electrically conductive. *Voyager 2* found that one of the most striking influences of the sideways position of the planet is its effect on the tail of the magnetic field, which is itself tilted 60 degrees from the planet's axis of rotation. The magnetotail was shown to be twisted by the planet's rotation into a long corkscrew shape behind the planet.

Radiation belts at Uranus were found to be of an intensity similar to those at Saturn. The intensity of radiation within the belts is such that irradiation would quickly darken (within 100,000 years) any methane trapped in the icy surfaces of the inner moons and ring particles. This may have contributed to the darkened surfaces of the moons and ring particles, which are almost uniformly gray in color.

A high layer of haze was detected around the sunlit pole, which also was found to radiate large amounts of ultraviolet light, a phenomenon dubbed "electroglow". The average temperature of the atmosphere of the planet is about 59 K (–214.2 °C). Surprisingly, the illuminated and dark poles, and most of the planet, show nearly the same temperature at the cloud tops.

Voyager 2 found 10 new moons, bringing the total number to 15 at the time. Most of the new moons are small, with the largest measuring about 150 km (93 mi) in diameter.

The moon Miranda, innermost of the five large moons, was revealed to be one of the strangest bodies yet seen in the Solar System. Detailed images from *Voyager 2*'s flyby of the moon showed huge oval structures termed *coronae* flanked by faults as deep as 20 km (12 mi), terraced layers, and a mixture of old and young surfaces. One theory holds that Miranda may be a reaggregation of material from an earlier time when the moon was fractured by a violent impact.

The five large moons appear to be ice–rock conglomerates like the satellites of Saturn. Titania is marked by huge fault systems and canyons indicating some degree of geologic, probably tectonic, activity in its history. Ariel has the brightest and possibly youngest surface of all the Uranian moons and also appears to have undergone geologic activity that led to many fault valleys and what seem to be extensive flows of icy material. Little geologic activity has occurred on Umbriel or Oberon, judging by their old and dark surfaces.

All nine previously known rings were studied by the spacecraft and showed the Uranian rings to be distinctly different from those at Jupiter and Saturn. The ring system may be relatively young and did not form at the same time as Uranus. Particles that make up the rings may be remnants of a moon that was broken by a high-velocity impact or torn up by gravitational effects. *Voyager 2* also discovered two new rings.

Proposed missions

Mission concepts to Uranus	Agency/country	Type
MUSE	ESA	orbiter and atmospheric probe
Oceanus	NASA/JPL	orbiter
ODINUS	ESA	twin orbiters
NASA Uranus orbiter and probe	NASA	orbiter and atmospheric probe
Uranus Pathfinder	United Kingdom	orbiter

Summary of missions to the outer Solar System

System Spacecraft	Jupiter	Saturn	Uranus	Neptune	Pluto
Pioneer 10	1973 flyby				
Pioneer 11	1974 flyby	1979 flyby			
Voyager 1	1979 flyby	1980 flyby			
Voyager 2	1979 flyby	1981 flyby	1986 flyby	1989 flyby	
Galileo	1995–2003 orbiter; 1995, 2003 atmospheric				
Ulysses	1992, 2004 gravity assist				
Cassini–Huygens	2000 gravity assist	2004–2017 orbiter; 2005 Titan lander			
New Horizons	2007 gravity assist				2015 flyby
Juno	2016– orbiter				
Jupiter Icy Moons Explorer	2022– Planned orbiter				
Europa Clipper	2025– Planned orbiter				

A number of missions to Uranus have been proposed. Scientists from the Mullard Space Science Laboratory in the United Kingdom have proposed the joint NASA–ESA *Uranus Pathfinder* mission to Uranus. A call for a medium-class (M-class) mission to the planet to be launched in 2022 was submitted to the ESA in December 2010 with the signatures of 120 scientists from across the globe. The ESA caps the cost of M-class missions at €470 million.[347]

In 2009, a team of planetary scientists from NASA's Jet Propulsion Laboratory advanced possible designs for a solar-powered Uranus orbiter. The most favorable launch window for such a probe would be in August 2018, with arrival at Uranus in September 2030. The science package may include magnetometers, particle detectors and, possibly, an imaging camera.[348]

In 2011, the United States National Research Council recommended a Uranus orbiter and probe as the third priority for a NASA Flagship mission by the NASA Planetary Science Decadal Survey. However, this mission is considered to be lower-priority than future missions to Mars and the Jovian System.[349]

A mission to Uranus is one of several proposed uses under consideration for the unmanned variant of NASA's heavy-lift Space Launch System (SLS) currently in development. The SLS would reportedly be capable of launching up to 1.7 metric tons to Uranus.

In 2013, it was proposed to use an electric sail (E-Sail) to send an atmospheric entry probe to Uranus.[350]

In 2015 NASA announced it had begun a feasibility study into the possibility of orbital missions to Uranus and Neptune, within a budget of $2 billion in 2015 dollars. According to NASA's planetary science director Jim Green, who initiated the study, such missions would launch in the late 2020s at the earliest, and would be contingent upon their endorsement by the planetary science community, as well as NASA's ability to provide nuclear power sources for the spacecraft. Conceptual designs for such a mission are currently being analyzed.[351]

MUSE, conceived in 2012 and proposed in 2015, is a European concept for a dedicated mission to the planet Uranus to study its atmosphere, interior, moons, rings, and magnetosphere. It is suggested to be launched with an Ariane 5 rocket in 2026, arriving at Uranus in 2044, and operating until 2050.

In 2016, another mission concept was conceived, called Origins and Composition of the Exoplanet Analog Uranus System (OCEANUS), and it was presented in 2017 as a potential contestant for a future New Frontiers program mission.[352]

Bibliography

* "Uranus Science Results"[353]. *Voyager Science Results at Uranus*. NASA. Retrieved February 27, 2013.
* Stone, E. C.; Miner, E. D. (1986). "The Voyager 2 Encounter with the Uranian System". *Science*. **233** (4759): 39–43. Bibcode: 1986Sci... 233...39S[354]. doi: 10.1126/science.233.4759.39[355]. PMID 17812888[356].
* Rick, Gore (1986). "Uranus Voyager visits a dark planet". *National Geographic*. **170** (2): 178–195. Bibcode: 1986NaGe..170..178G[357].

External links

* NASA *Voyager* website[358]

Appendix

References

[1] These are the mean elements from VSOP87, together with derived quantities.

[2] Calculated using data from Seidelmann, 2007. UNIQ-ref-0-cf064b24719039b5-QINU

[3] Calculation of He, H_2 and CH_4 molar fractions is based on a 2.3% mixing ratio of methane to hydrogen and the 15/85 He/H_2 proportions measured at the tropopause.

[4] Journal of the Royal Society and Royal Astronomical Society 1, 30, quoted in Miner, p. 8.

[5] Royal Astronomical Society MSS W.2/1.2, 23; quoted in Miner p. 8.

[6] RAS MSS Herschel W.2/1.2, 24, quoted in Miner p. 8.

[7] RAS MSS Herschel W1/13.M, 14 quoted in Miner p. 8.

[8] Johann Elert Bode, Berliner Astronomisches Jahrbuch, p. 210, 1781, quoted in Miner, p. 11.

[9] Miner, p. 11.

[10] Because, in the English-speaking world, the latter sounds like "your anus", the former pronunciation also saves embarrassment: as Pamela Gay, an astronomer at Southern Illinois University Edwardsville, noted on her podcast, to avoid "being made fun of by any small schoolchildren ... when in doubt, don't emphasise anything and just say /ˈjʊərənəs/. And then run, quickly."<ref>

[11] RAS MSS Herschel W.1/12.M, 20, quoted in Miner, p. 12

[12]: [In original German]: "Bereits in der am 12ten März 1782 bei der hiesigen naturforschenden Gesellschaft vorgelesenen Abhandlung, habe ich den Namen des Vaters vom Saturn, nemlich Uranos, oder wie er mit der lateinischen Endung gewöhnlicher ist, Uranus vorgeschlagen, und habe seit dem das Vergnügen gehabt, daß verschiedene Astronomen und Mathematiker in ihren Schriften oder in Briefen an mich, diese Benennung aufgenommen oder gebilligt. Meines Erachtens muß man bei dieser Wahl die Mythologie befolgen, aus welcher die uralten Namen der übrigen Planeten entlehnen worden; denn in der Reihe der bisher bekannten, würde der von einer merkwürdigen Person oder Begebenheit der neuern Zeit wahrgenommene Name eines Planeten sehr auffallen. Diodor von Cicilien erzahlt die Geschichte der Atlanten, eines uralten Volks, welches eine der fruchtbarsten Gegenden in Africa bewohnte, und die Meeresküsten seines Landes als das Vaterland der Götter ansah. Uranus war ihr, erster König, Stifter ihres gesitteter Lebens und Erfinder vieler nützlichen Künste. Zugleich wird er auch als ein fleißiger und geschickter Himmelsforscher des Alterthums beschrieben... Noch mehr: Uranus war der Vater des Saturns und des Atlas, so wie der erstere der Vater des Jupiters."; [Translated]: "Already in the pre-read at the local Natural History Society on 12th March 1782 treatise, I have the father's name from Saturn, namely Uranus, or as it is usually with the Latin suffix, proposed Uranus, and have since had the pleasure that various astronomers and mathematicians, cited in their writings or letters to me approving this designation. In my view, it is necessary to follow the mythology in this election, which had been borrowed from the ancient name of the other planets; because in the series of previously known, perceived by a strange person or event of modern times name of a planet would very noticeable. Diodorus of Cilicia tells the story of Atlas, an ancient people that inhabited one of the most fertile areas in Africa, and looked at the sea shores of his country as the homeland of the gods. Uranus was her first king, founder of their civilized life and inventor of many useful arts. At the same time he is also described as a diligent and skilful astronomers of antiquity ... even more: Uranus was the father of Saturn and the Atlas, as the former is the father of Jupiter."

[13] Jean Meeus, *Astronomical Algorithms* (Richmond, VA: Willmann-Bell, 1998) p 271. From the 1841 aphelion to the 2092 one, perihelia are always 18.28 and aphelia always 20.10 astronomical units

[14] Mixing ratio is defined as the number of molecules of a compound per a molecule of hydrogen.

[15] Michael Schirber – **Missions Proposed to Explore Mysterious Tilted Planet Uranus** (2011) – Astrobiology Magazine http://www.space.com/13248-nasa-uranus-missions-solar-system. html. Space.com. Retrieved on 2 April 2012.

[16] https://books.google.com/books?id=ZqA5AAAAcAAJ

[17] //www.worldcat.org/issn/0027-9358

[18] //www.worldcat.org/oclc/643483454

[19] https://www.britannica.com/EBchecked/topic/619284

[20] http://sci.esa.int/science-e/www/object/index.cfm?fobjectid=35653

[21] http://nssdc.gsfc.nasa.gov/planetary/factsheet/uranusfact.html

[22] https://web.archive.org/web/20070624113641/http://solarsystem.nasa.gov/planets/profile.cfm?Object=Uranus

[23] http://solarsystem.nasa.gov/

[24] http://www.projectshum.org/Planets/uranus.html

[25] http://photojournal.jpl.nasa.gov/targetFamily/Uranus

[26] http://www.ciclops.org/ir_index/81/Voyager_at_Uranus

[27] http://www.astronomycast.com/astronomy/episode-62-uranus/

[28] http://www.solarviews.com/raw/uranus/urfamily.jpg

[29] http://www.sixtysymbols.com/videos/uranus.htm

[30] https://www.youtube.com/watch?v=h3ppbbYXMxE

[31] Lunine 1993, pp. 219–222.

[32] de Pater Romani et al. 1991, p. 231, Fig. 13.

[33] Fegley Gautier et al. 1991, pp. 151–154.

[34] Lockyer 1889.

[35] Huggins 1889.

[36] Adel & Slipher 1934.

[37] Kuiper 1949.

[38] Herzberg 1952.

[39] Pearl Conrath et al. 1990, pp. 12–13, Table I.

[40] Smith 1984, pp. 213–214.

[41] Stone 1987, p. 14,874, Table 3.

[42] Fegley Gautier et al. 1991, pp. 155–158, 168–169.

[43] Smith Soderblom et al. 1986, pp. 43–49.

[44] Sromovsky & Fry 2005, pp. 459–460.

[45] Sromovsky & Fry 2005, p. 469, Fig.5.

[46] Lunine 1993, pp. 222–230.

[47] Tyler Sweetnam et al. 1986, pp. 80–81.

[48] Conrath Gautier et al. 1987, p. 15,007, Table 1.

[49] Lodders 2003, pp. 1,228–1,230.

[50] Conrath Gautier et al. 1987, pp. 15,008–15,009.

[51] NASA NSSDC, Uranus Fact Sheet http://nssdc.gsfc.nasa.gov/planetary/factsheet/uranusfact.html (retrieved 7 Oc 2015)

[52] Lunine 1993, pp. 235–240.

[53] Lindal Lyons et al. 1987, pp. 14,987, 14,994–14,996.

[54] Bishop Atreya et al. 1990, pp. 457–462.

[55] Atreya & Wong 2005, pp. 130–131.

[56] de Pater Romani et al. 1989, pp. 310–311.

[57] Encrenaz 2005, pp. 107–110.

[58] Encrenaz 2003, pp. 98–100, Table 2 on p. 96.

[59] Feuchtgruber Lellouch et al. 1999.

[60] Burgdorf Orton et al. 2006, pp. 634–635.

[61] Bishop Atreya et al. 1990, p. 448.

[62] Summers & Strobel 1989, pp. 496–497.

[63] Encrenaz 2003, p. 93.

[64] Burgdorf Orton et al. 2006, p. 636.

[65] Encrenaz 2003, p. 92.

[66] Encrenaz Lellouch et al. 2004, p. L8.

[67]

[68] Herbert Sandel et al. 1987, p. 15,097, Fig. 4.

[69] Lunine 1993, pp. 240–245.

[70] Hanel Conrath et al. 1986, p. 73.

[71] Pearl Conrath et al. 1990, p. 26, Table IX.

[72] Sromovsky Irwin et al. 2006, pp. 591–592.

[73] Sromovsky Irwin et al. 2006, pp. 592–593.

[74] Indeed, a recent analysis based on a new data set of the methane absorption coefficients shifted the clouds to 1.6 and 3 bar, respectively.<ref name="FOOTNOTEFrySromovsky2009">Fry & Sromovsky 2009.

[75] Irwin Teanby et al. 2010, p. 913.

[76] Irwin Teanby et al. 2007, pp. L72–L73.

[77] Sromovsky & Fry 2005, p. 483.

[78] Hammel Sromovsky et al. 2009, p. 257.

[79] Hammel & Lockwood 2007, pp. 291–293.

[80] Herbert Sandel et al. 1987, pp. 15,101–15,102.

[81] Lunine 1993, pp. 230–234.

[82] Young 2001, pp. 241–242.

[83] Summers & Strobel 1989, pp. 497, 502, Fig. 5a.

[84] Herbert & Sandel 1999, pp. 1,123–1,124.

[85] In 1986 the stratosphere was poorer in hydrocarbons at the poles than near the equator;<ref name="FOOTNOTEBishop Atreya et al.1990457–462">Bishop Atreya et al. 1990, pp. 457–462.

[86] Herbert & Sandel 1999, pp. 1,130–1,131.

[87] Young 2001, pp. 239–240, Fig. 5.

[88] Encrenaz 2005, p. 111, Table IV.

[89] Pollack Rages et al. 1987, p. 15,037.

[90] Lunine 1993, p. 229, Fig. 3.

[91] At these altitudes the temperature has local maxima, which may be caused by absorption of solar radiation by haze particles.<ref name="FOOTNOTELunine1993222–230">Lunine 1993, pp. 222–230.

[92] Bishop Atreya et al. 1990, pp. 462–463.

[93] Smith Soderblom et al. 1986, pp. 43–46.

[94] Herbert & Sandel 1999, pp. 1,122–1,123.

[95] Miller Aylward et al. 2005, p. 322, Table I.

[96] Herbert Sandel et al. 1987, pp. 15,107–15,108.

[97] Tyler Sweetnam et al. 1986, p. 81.

[98] Lindal Lyons et al. 1987, p. 14,992, Fig. 7.

[99] Trafton Miller et al. 1999, pp. 1,076–1,078.

[100] Encrenaz Drossart et al. 2003, pp. 1,015–1,016.

[101] The total power input into the aurora is 3–7×10^{10} W—insufficient to heat up the thermosphere.<ref name="FOOTNOTEHerbertSandel19991,133–1,135">Herbert & Sandel 1999, pp. 1,133–1,135.

[102] Herbert & Sandel 1999, pp. 1,133–1,135.

[103] Lam Miller et al. 1997, pp. L75–76.

[104] The hot thermosphere of Uranus produces hydrogen quadrupole emission lines in the near-infrared part of the spectrum (1.8–2.5 μm) with the total emitted power of 1–2×10^{10} W. The power emitted by molecular hydrogen in the far infrared part of the spectrum is about 2×10^{11} W.<ref name="FOOTNOTETrafton Miller et al.19991,073–1,076">Trafton Miller et al. 1999, pp. 1,073–1,076.

[105] Trafton Miller et al. 1999, pp. 1,073–1,076.

[106] Miller Achilleos et al. 2000, pp. 2,496–2,497.

[107] Herbert & Sandel 1999, pp. 1,127–1,128, 1,130–1,131.

[108] The scale height sh is defined as $sh = RT/(Mg_j)$, where $R = 8.31$ J/mol/K is the gas constant, $M \approx 0.0023$ kg/mol is the average molar mass in the Uranian atmosphere,<ref name="FOOTNOTELunine1993222–230">Lunine 1993, pp. 222–230.

[109] Herbert & Hall 1996, p. 10,877.

[110] Herbert & Hall 1996, p. 10,879, Fig. 2.

[111] Herbert & Sandel 1999, p. 1,124.

[112] Herbert Sandel et al. 1987, pp. 15,102–15,104.

[113] The corona contains a significant population of supra-thermal (energy of up to 2 eV) hydrogen atoms. Their origin is unclear, but they may be produced by the same mechanism that heats the thermosphere.<ref name="FOOTNOTEHerbertHall199610,880–10,882">Herbert & Hall 1996, pp. 10,880–10,882.

[114] Herbert & Hall 1996, pp. 10,879–10,880.

[115] Rages Hammel et al. 2004, p. 548.

[116] Sromovsky & Fry 2005, pp. 470–472, 483, Table 7, Fig. 6.

[117] Sromovsky Fry et al. 2009, p. 265.

[118] Sromovsky & Fry 2005, pp. 474–482.

[119] Smith Soderblom et al. 1986, pp. 47–49.

[120] Hammel & Lockwood 2007, pp. 293–296.

[121] http://adsabs.harvard.edu/abs/1934PhRv...46..902A

[122] //doi.org/10.1103/PhysRev.46.902

[123] http://www-personal.umich.edu/~atreya/Chapters/2005_JovianCloud_Multiprobes.pdf

[124] http://adsabs.harvard.edu/abs/2005SSRv..116..121A

[125] //doi.org/10.1007/s11214-005-1951-5

[126] http://www-personal.umich.edu/~atreya/Articles/1990_Reanalysis.pdf

[127] http://adsabs.harvard.edu/abs/1990Icar...88..448B

[128] //doi.org/10.1016/0019-1035%2890%2990094-P

[129] http://adsabs.harvard.edu/abs/2006Icar..184..634B

[130] //doi.org/10.1016/j.icarus.2006.06.006

[131] http://adsabs.harvard.edu/abs/1987JGR....9215003C

[132] //doi.org/10.1029/JA092iA13p15003

[133] http://adsabs.harvard.edu/abs/2003P&SS...51...89E

[134] //doi.org/10.1016/S0032-0633%2802%2900145-9

[135] http://www-personal.umich.edu/~atreya/Articles/2003_Rotational_Temperature.pdf

[136] http://adsabs.harvard.edu/abs/2003P&SS...51.1013E

[137] //doi.org/10.1016/j.pss.2003.05.010

[138] http://www-personal.umich.edu/~atreya/Articles/2004_First_Detection.pdf

[139] http://adsabs.harvard.edu/abs/2004A&A...413L...5E

[140] //doi.org/10.1051/0004-6361%3A20034637

[141] http://adsabs.harvard.edu/abs/2005SSRv..116...99E

[142] //doi.org/10.1007/s11214-005-1950-6

[143] http://solarsystem.wustl.edu/wp-content/uploads/reprints/1991/No39%20Fegley%20et%20al%201991%20Uranus.pdf

[144] //www.worldcat.org/oclc/22625114

[145] http://adsabs.harvard.edu/abs/1999A&A...341L..17F

[146] http://adsabs.harvard.edu/abs/2009DPS....41.1406F

[147] http://adsabs.harvard.edu/abs/2007Icar..186..291H

[148] //doi.org/10.1016/j.icarus.2006.08.027

[149] https://web.archive.org/web/20110719231643/http://epicwiki.atmos.louisville.edu/images/Hammel09.pdf

[150] http://adsabs.harvard.edu/abs/2009Icar..201..257H

[151] //doi.org/10.1016/j.icarus.2008.08.019

[152] http://epicwiki.atmos.louisville.edu/images/Hammel09.pdf

[153] http://adsabs.harvard.edu/abs/1986Sci...233...70H

[154] //doi.org/10.1126/science.233.4759.70

[155] //www.ncbi.nlm.nih.gov/pubmed/17812891

[156] http://www-personal.umich.edu/~atreya/Articles/1987_Upper_Atm_Uranus.pdf

[157] http://adsabs.harvard.edu/abs/1987JGR....9215093H

[158] //doi.org/10.1029/JA092iA13p15093

[159] http://adsabs.harvard.edu/abs/1996JGR...10110877H

[160] //doi.org/10.1029/96JA00427

[161] http://adsabs.harvard.edu/abs/1999P&SS...47.1119H

[162] //doi.org/10.1016/S0032-0633%2898%2900142-1

[163] http://adsabs.harvard.edu/abs/1952ApJ...115..337H

[164] //doi.org/10.1086/145552
[165] http://adsabs.harvard.edu/abs/1889MNRAS..49Q.404H
[166] //doi.org/10.1093/mnras/49.8.403a
[167] http://adsabs.harvard.edu/abs/2007ApJ...665L..71I
[168] //doi.org/10.1086/521189
[169] http://adsabs.harvard.edu/abs/2010Icar..208..913I
[170] //doi.org/10.1016/j.icarus.2010.03.017
[171] http://adsabs.harvard.edu/abs/1949ApJ...109..540K
[172] //doi.org/10.1086/145161
[173] http://www.ucl.ac.uk/phys/amopp/people/jonathan_tennyson/papers/192.pdf
[174] http://adsabs.harvard.edu/abs/1997ApJ...474L..73L
[175] //doi.org/10.1086/310424
[176] http://adsabs.harvard.edu/abs/1987JGR....9214987L
[177] //doi.org/10.1029/JA092iA13p14987
[178] http://adsabs.harvard.edu/abs/1889AN....121..369L
[179] //doi.org/10.1002/asna.18891212402
[180] http://weft.astro.washington.edu/courses/astro557/LODDERS.pdf
[181] http://adsabs.harvard.edu/abs/2003ApJ...591.1220L
[182] //doi.org/10.1086/375492
[183] http://adsabs.harvard.edu/abs/1993ARA&A..31..217L
[184] //doi.org/10.1146/annurev.aa.31.090193.001245
[185] http://www.ucl.ac.uk/~ucaptss/work/publications/royalsoc/royalsoc.pdf
[186] //doi.org/10.1098/rsta.2000.0662
[187] http://adsabs.harvard.edu/abs/2005SSRv..116..319M
[188] //doi.org/10.1007/s11214-005-1960-4
[189] http://adsabs.harvard.edu/abs/2004Icar..172..548R
[190] //doi.org/10.1016/j.icarus.2004.07.009
[191] http://www-personal.umich.edu/~atreya/Articles/1989_Uranus_Deep_Atm.pdf
[192] http://adsabs.harvard.edu/abs/1989Icar...82..288D
[193] //doi.org/10.1016/0019-1035%2889%2990040-7
[194] http://www-personal.umich.edu/~atreya/Articles/1991_Microwave_Absorption.pdf
[195] http://adsabs.harvard.edu/abs/1991Icar...91..220D
[196] //doi.org/10.1016/0019-1035%2891%2990020-T
[197] http://adsabs.harvard.edu/abs/1990Icar...84...12P
[198] //doi.org/10.1016/0019-1035%2890%2990155-3
[199] http://www-personal.umich.edu/~atreya/Articles/1987_Stratospheric_Haze.pdf
[200] http://adsabs.harvard.edu/abs/1987JGR....9215037P
[201] //doi.org/10.1029/JA092iA13p15037
[202] http://adsabs.harvard.edu/abs/1984NASCP2330..213S
[203] http://adsabs.harvard.edu/abs/1986Sci...233...43S
[204] //doi.org/10.1126/science.233.4759.43
[205] //www.ncbi.nlm.nih.gov/pubmed/17812889
[206] //arxiv.org/abs/1503.03714
[207] http://adsabs.harvard.edu/abs/2005Icar..179..459S
[208] //doi.org/10.1016/j.icarus.2005.07.022
[209] http://adsabs.harvard.edu/abs/2006Icar..182..577S
[210] //doi.org/10.1016/j.icarus.2006.01.008
[211] //arxiv.org/abs/1503.01957
[212] http://adsabs.harvard.edu/abs/2009Icar..203..265S
[213] //doi.org/10.1016/j.icarus.2009.04.015
[214] http://adsabs.harvard.edu/abs/1989ApJ...346..495S
[215] //doi.org/10.1086/168031
[216] http://adsabs.harvard.edu/abs/1987JGR....9214873S
[217] //doi.org/10.1029/JA092iA13p14873
[218] http://adsabs.harvard.edu/abs/1999ApJ...524.1059T
[219] //doi.org/10.1086/307838

[220] http://adsabs.harvard.edu/abs/1986Sci...233...79T
[221] //doi.org/10.1126/science.233.4759.79
[222] //www.ncbi.nlm.nih.gov/pubmed/17812893
[223] http://www.boulder.swri.edu/~layoung/eprint/ur149/Young2001Uranus.pdf
[224] http://adsabs.harvard.edu/abs/2001Icar..153..236Y
[225] //doi.org/10.1006/icar.2001.6698
[226] Sromovsky & Fry 2005.
[227] Soderblom et al. 1986.
[228] Lakdawalla 2004.
[229] Hammel Sromovsky et al. 2009.
[230] Smith Soderblom et al. 1986.
[231] Hammel de Pater et al. ("Uranus in 2003") 2005.
[232] Rages Hammel et al. 2004.
[233] Sromovsky Fry et al. 2009.
[234] Karkoschka ("Uranus") 2001.
[235] Hammel de Pater et al. ("Uranus in 2004") 2005.
[236] Sromovsky Fry et al. 2006.
[237] Hanel Conrath et al. 1986.
[238] Hammel Rages et al. 2001.
[239] Lockwood & Jerzykiewicz 2006.
[240] Klein & Hofstadter 2006.
[241] Young 2001.
[242] Hofstadter & Butler 2003.
[243] Hammel & Lockwood 2007.
[244] Devitt 2004.
[245] Pearl Conrath et al. 1990.
[246] Lunine 1993.
[247] Podolak Weizman et al. 1995.
[248] http://www.news.wisc.edu/10402
[249] http://www.llnl.gov/tid/lof/documents/pdf/316112.pdf
[250] http://adsabs.harvard.edu/abs/2005Icar..175..534H
[251] //doi.org/10.1016/j.icarus.2004.11.012
[252] http://www.llnl.gov/tid/lof/documents/pdf/316113.pdf
[253] http://adsabs.harvard.edu/abs/2005Icar..175..284H
[254] //doi.org/10.1016/j.icarus.2004.11.016
[255] http://adsabs.harvard.edu/abs/2007Icar..186..291H
[256] //doi.org/10.1016/j.icarus.2006.08.027
[257] http://adsabs.harvard.edu/abs/2001Icar..153..229H
[258] //doi.org/10.1006/icar.2001.6689
[259] https://web.archive.org/web/20110719231643/http://epicwiki.atmos.louisville.edu/images/
Hammel09.pdf
[260] http://adsabs.harvard.edu/abs/2009Icar..201..257H
[261] //doi.org/10.1016/j.icarus.2008.08.019
[262] http://epicwiki.atmos.louisville.edu/images/Hammel09.pdf
[263] http://adsabs.harvard.edu/abs/1986Sci...233...70H
[264] //doi.org/10.1126/science.233.4759.70
[265] //www.ncbi.nlm.nih.gov/pubmed/17812891
[266] http://adsabs.harvard.edu/abs/2003Icar..165..168H
[267] //doi.org/10.1016/S0019-1035%2803%2900174-X
[268] http://adsabs.harvard.edu/abs/2001Icar..151...84K
[269] //doi.org/10.1006/icar.2001.6599
[270] http://adsabs.harvard.edu/abs/2006Icar..184..170K
[271] //doi.org/10.1016/j.icarus.2006.04.012
[272] http://www.planetary.org/news/2004/1111_No_Longer_Boring_Fireworks_and_Other.html
[273] http://adsabs.harvard.edu/abs/2006Icar..180..442L
[274] //doi.org/10.1016/j.icarus.2005.09.009

[275] http://adsabs.harvard.edu/abs/1993ARA&A..31..217L
[276] //doi.org/10.1146/annurev.aa.31.090193.001245
[277] http://adsabs.harvard.edu/abs/1990Icar...84...12P
[278] //doi.org/10.1016/0019-1035%2890%2990155-3
[279] //www.worldcat.org/issn/0019-1035
[280] http://adsabs.harvard.edu/abs/1995P&SS...43.1517P
[281] //doi.org/10.1016/0032-0633%2895%2900061-5
[282] http://adsabs.harvard.edu/abs/2004Icar..172..548R
[283] //doi.org/10.1016/j.icarus.2004.07.009
[284] http://adsabs.harvard.edu/abs/1986Sci...233...43S
[285] //doi.org/10.1126/science.233.4759.43
[286] //www.ncbi.nlm.nih.gov/pubmed/17812889
[287] //arxiv.org/abs/1503.03714
[288] http://adsabs.harvard.edu/abs/2005Icar..179..459S
[289] //doi.org/10.1016/j.icarus.2005.07.022
[290] //arxiv.org/abs/1503.01957
[291] http://adsabs.harvard.edu/abs/2009Icar..203..265S
[292] //doi.org/10.1016/j.icarus.2009.04.015
[293] http://www.physorg.com/pdf78676690.pdf
[294] http://www.boulder.swri.edu/~layoung/eprint/url49/Young2001Uranus.pdf
[295] http://adsabs.harvard.edu/abs/2001Icar..153..236Y
[296] //doi.org/10.1006/icar.2001.6698
[297] http://www.space.com/18707-uranus-temperature.html
[298] https://web.archive.org/web/20141129055142/http://interesting-facts.com/uranus-facts/
[299] An astronomical unit, or AU, is the average distance between the Earth and the Sun, or about 150 million kilometres. It is the standard unit of measurement for interplanetary distances.
[300] Zeilik & Gregory 1998, p. 207.
[301] The combined mass of Jupiter, Saturn, Uranus and Neptune is 445.6 Earth masses. The mass of remaining material is ~5.26 Earth masses or 1.1% (see Solar System#Notes and List of Solar System objects by mass)
[302] The reason that Saturn, Uranus and Neptune all moved outward whereas Jupiter moved inward is that Jupiter is massive enough to eject planetesimals from the Solar System, while the other three outer planets are not. To eject an object from the Solar System, Jupiter transfers energy to it, and so loses some of its own orbital energy and moves inwards. When Neptune, Uranus and Saturn perturb planetesimals outwards, those planetesimals end up in highly eccentric but still bound orbits, and so can return to the perturbing planet and possibly return its lost energy. On the other hand, when Neptune, Uranus and Saturn perturb objects inwards, those planets gain energy by doing so and therefore move outwards. More importantly, an object being perturbed inwards stands a greater chance of encountering Jupiter and being ejected from the Solar System, in which case the energy gains of Neptune, Uranus and Saturn obtained from their inwards deflections of the ejected object become permanent.
[303]

See also
[304] Zeilik & Gregory 1998, pp. 118–120.
[305] In all of these cases of transfer of angular momentum and energy, the angular momentum of the two-body system is conserved. In contrast, the summed energy of the moon's revolution plus the primary's rotation is not conserved, but decreases over time, due to dissipation via frictional heat generated by the movement of the tidal bulge through the body of the primary. If the primary were a frictionless ideal fluid, the tidal bulge would be centered under the satellite, and no transfer would take place. It is the loss of dynamical energy through friction that makes transfer of angular momentum possible.
[306] Duncan & Lissauer 1997.
[307] Zeilik & Gregory 1998, p. 320–321.
[308] Zeilik & Gregory 1998, p. 322.
[309] http://adsabs.harvard.edu/abs/1997Icar..125....1D
[310] //doi.org/10.1006/icar.1996.5568

[311] http://arquivo.pt/wayback/20160520023943/http://media.skyandtelescope.com/video/Solar_System_Sim.mov

[312] http://www.skyandtelescope.com

[313] http://www.cfa.harvard.edu/seuforum/animations/animations/galaxycollision.mpg

[314] http://www.space.com/common/media/video/player.php?videoRef=mm32_SunDeath

[315] The mass of Triton is about 2.14×10^{22} kg, whereas the combined mass of the Uranian moons is about 0.92×10^{22} kg.

[316] Uranus mass of 8.681×10^{25} kg / Mass of Uranian moons of 0.93×10^{22} kg

[317] The axial tilt of Uranus is 97°.

[318] Order refers to the position among other moons with respect to their average distance from Uranus.

[319] Label refers to the Roman numeral attributed to each moon in order of their discovery.

[320] Diameters with multiple entries such as "60 $\times$ 40 $\times$ 34" reflect that the body is not a perfect spheroid and that each of its dimensions have been measured well enough. The diameters and dimensions of Miranda, Ariel, Umbriel, and Oberon were taken from Thomas, 1988. The diameter of Titania is from Widemann, 2009. The dimensions and radii of the inner moons are from Karkoschka, 2001, except for Cupid and Mab, which were taken from Showalter, 2006. The radii of outer moons except Sycorax were taken from Sheppard, 2005. The diameter of Sycorax is from Lellouch, 2013.

[321] Masses of Miranda, Ariel, Umbriel, Titania, and Oberon were taken from Jacobson, 1992. Masses of all other moons were calculated assuming a density of 1.3 g/cm^3 and using given radii.

[322] Negative orbital periods indicate a retrograde orbit around Uranus (opposite to the planet's rotation).

[323] Inclination measures the angle between the moon's orbital plane and the plane defined by Uranus's equator.

[324] Detected in 2001, published in 2003.

[325] https://web.archive.org/web/20110823202755/http://orinetz.com/planet/tourprog/uranusmoons.html

[326] http://solarsystem.nasa.gov/planets/uranus/moons

[327] http://hubblesite.org/news_release/news/2005-33

[328] http://home.dtm.ciw.edu/users/sheppard/satellites/urasatdata.html

[329] https://planetarynames.wr.usgs.gov/?System=Uranus

[330] (re study by Stuart Eves)

[331] Forward-scattered light is the light scattered at a small angle relative to the solar light (phase angle close to 180°).

[332] *Off opposition* means that the angle between the object-sun direction and object-Earth direction is not zero.

[333] The normal optical depth τ of a ring is the ratio of the total geometrical cross-section of the ring's particles to the square area of the ring. It assumes values from zero to infinity. A light beam passing normally through a ring will be attenuated by the factor $e^{-\tau}$.(re study by Stuart Eves)

[334] The equivalent depth ED of a ring is defined as an integral of the normal optical depth across the ring. In other words UNIQ-nowiki-1-cf064b24719039b5-QINU , where r is radius.

[335] Back-scattered light is the light scattered at an angle close to 180° relative to the solar light (phase angle close to 0°).

[336] The radii of the 6,5,4, α, β, η, γ, δ, λ and ϵ rings were taken from Esposito et al., 2002. The widths of the 6,5,4, α, β, η, γ, δ and ϵ rings are from Karkoschka et al., 2001. The radii and widths of the ζ and 1986U2R rings were taken from de Pater et al., 2006. The width of the λ ring is from Holberg et al., 1987. The radii and widths of the μ and ν rings were extracted from Showalter et al., 2006.

[337] The equivalent depth of the 1986U2R and ζ_c/ζ_{cc} rings is a product of their width and the normal optical depth. The equivalent depths of the 6,5,4, α, β, η, γ, δ and ϵ rings were taken from Karkoschka et al., 2001. The equivalent depths of the λ and ζ, μ and ν rings are derived using μEW values from de Pater et al., 2006 and de Pater et al., 2006b, respectively. The μEW values for these rings were multiplied by the factor 20, which corresponds to the assumed albedo of the ring's particles of 5%.

[338] The normal optical depths of all rings except ζ, ζ_c, ζ_{cc}, 1986U2R, μ and ν were calculated as ratios of the equivalent depths to the widths. The normal optical depth of the 1986U2R ring was taken from de Smith et al., 1986. The normal optical depths of the μ and ν rings are peak values from Showalter et al., 2006, while the normal optical depths of ζ, ζ_c and ζ_{cc} rings are from Dunn eta al., 2010.

[339] The thickness estimates are from Lane et al., 1986.

[340] The rings' eccentricities and inclinations were taken from Stone et al., 1986 and French et al., 1989.<ref name=French1988>

[341] https://web.archive.org/web/20070609080249/http://solarsystem.nasa.gov/planets/profile. cfm?Object=Uranus&Display=Rings

[342] http://solarsystem.nasa.gov

[343] http://nssdc.gsfc.nasa.gov/planetary/factsheet/uranringfact.html

[344] http://hubblesite.org/newscenter/newsdesk/archive/releases/2005/33/

[345] http://planetarynames.wr.usgs.gov/append8.html

[346] //en.wikipedia.org/w/index.php?title=Exploration_of_Uranus&action=edit

[347] ESA official website: "Call for a Medium-size mission opportunity for a launch in 2022" http: //sci.esa.int/science-e/www/object/index.cfm?fobjectid=47570. January 16, 2011. Retrieved January 16, 2011.

[348] See also a draft http://www.lpi.usra.edu/decadal/opag/UranusOrbiter_v7.pdf.

[349] Mark Hofstadter, "Ice Giant Science: The Case for a Uranus Orbiter" http://www. spacepolicyonline.com/images/stories/PSDS%20GP1%20Hofstadter_Uranus%20Orbiter.pdf, Jet Propulsion Laboratory/California Institute of Technology, *Report to the Decadal Survey Giant Planets Panel*, 24 August 2009

[350] Fast E-sail Uranus entry probe mission https://arxiv.org/abs/1312.6554

[351] Stephen Clark "Uranus, Neptune in NASA's sights for new robotic mission" http:// spaceflightnow.com/2015/08/25/uranus-neptune-in-nasas-sights-for-new-robotic-mission/, *Spaceflight Now,* August 25, 2015

[352] New Frontiers-Class Missions to the Ice Giants https://www.hou.usra.edu/meetings/V2050/ pdf/8147.pdf. C. M. Elder, A. M. Bramson, L. W. Blum, H. T. Chilton, A. Chopra, C. Chu6, A. Das, A. Davis, A. Delgado, J. Fulton, L. Jozwiak, A. Khayat, M. E. Landis, J. L. Molaro, M. Slipski, S. Valencia11, J. Watkins, C. L. Young, C. J. Budney, K. L. Mitchell. Planetary Science Vision 2050 Workshop 2017 (LPI Contrib. No. 1989).

[353] http://voyager.jpl.nasa.gov/science/uranus.html

[354] http://adsabs.harvard.edu/abs/1986Sci...233...39S

[355] //doi.org/10.1126/science.233.4759.39

[356] //www.ncbi.nlm.nih.gov/pubmed/17812888

[357] http://adsabs.harvard.edu/abs/1986NaGe..170..178G

[358] http://voyager.jpl.nasa.gov

Article Sources and Contributors

The sources listed for each article provide more detailed licensing information including the copyright status, the copyright owner, and the license conditions.

Uranus *Source:* https://en.wikipedia.org/w/index.php?oldid=855736510 *License:* Creative Commons Attribution-Share Alike 3.0 *Contributors:* "alyosha", 1Martin33, 564dude, A. Parrot, A.amitkumar, A2soup, Abductive, Abishai 300, Acopyeditor, Albanaco, AlbertBickford, Alsee, Ambi Valent, Andrew Gray, Andyjsmith, Anthony22, ArtGriggs, AstroLynx, Azcolvin429, B14709, BD2412, Barjimoa, BatteryIncluded, Beauty School Dropout, Bender235, Blobbie244, Bob rulz, Boltz6100, CV9933, Cancina5645, Chackerian, Chaheel Riens, Cheetahrock21, ChieftanTartarus, Chivas112, Clockery, ClueBot NG, Coinmanj, Corinne, CouncilConnect, Coviekiller5, DN-boards1, DOSGuy, Dcirovic, Deejaye6, Deepanshu1707, Deturtlemon1, Dilidor, Dinkytown, Dja1979, Dominicanpapi82, Donutcity, Dotdotdotatsignapostrophe, Double sharp, DrKay, Drbogdan, EP111, Earthandmoon, Eggishorn, Egsan Bacon, Ehrenkater, Epicgenius, EricAnthony101, Ewen, Eyreland, Fartherred, FeRDNYC, FerJox, Finell, Finnusertop, FlightTime, Florian Blaschke, Fourthords, Frietjes, Gallina3795, Gob Lofa, Gparker, HalloweenNight, Hamiltpj, Headbomb, Hellbus, Hillbillyholiday, Huntster, Iggy the Swan, Iph, Isambard Kingdom, Jcpag2012, Jguy, Jmencisom, JoeHebda, Joeleoj123, John, Johnuniq, Jomey, Jon Kolbert, Jonathan Barrett, Jonesey95, JorisvS, JustAMuggle, JustinTimeCuber, KAP03, Katydidit, Kheider, Koavf, Kwamikagami, Kyucasio, LahmacunKebab, LandruBek, Lentower, Lindenhurst Liberty, Lithopsian, Logeion, Loraof, Magyar25, MartinZ, Matanbz, Materialscientist, Mean as custard, Mikhail Ryazanov, Moonraker12, Mortense, Mr Stephen, N Shar, Naddy, Nardog, NebY, Nedim Ardoğa, Newone, Niceguyedc, Nmillerche, Nudelkopp17, Nvvchar, ONUnicorn, OccultZone, Oldag07, Onceinawhile, Onore Baka Sama, Orange-kun, Ospalh, Paintspot, Pdebee, PhilipTerryGraham, Plagktos, Plastikspork, PlayStation 14, Primergrey, PulauKakatua19, Q6637p, Rab V, RagingR2, Randy Kryn, Reatlas, Rfassbind, Rhlozier, Richard L. Peterson, Rivertorch, Rjwilmsi, Robert the Devil, Robinh, RockMagnetist (DCO visiting scholar), Rondo66, Rothorpe, Ruslik0, Ryulong, Sailsbystars, SamX, Santa Claus of the Future, Saros136, Serendipodous, SheikhTheBaby, Shenme, SkateTier, SkoreKeep, Smk65536, Sol2y, SoylentPurple, Spartanwolf223, Spiderjerky, Srich32977, Starcluster, StringTheory11, TAnthony, Teddytktchan, Tetra quark, Thanhtinsaosang, The Rambling Man, The Transhumanist, Thedrumsmasher, Thewriter6669, Tim!, Toa Nidhiki05, Tom.Reding, Tomeasy, TommyBoy, Tomruen, Trackteur, Transphasic, Unbuttered Parsnip, Voello, Vsmith, W like wiki, Wavyinfinity, Whoop whoop pull up, WolfmanSF, Wrxsti fan, ZarhanFastfire, חסרשם, 展翅飛翔 1

Atmosphere of Uranus *Source:* https://en.wikipedia.org/w/index.php?oldid=843084507 *License:* Creative Commons Attribution-Share Alike 3.0 *Contributors:* A. Parrot, AManWithNoPlan, Acroterion, Alansohn, BatteryIncluded, Bluefist, Bobamnertiopsis, Bped1985, Brianga, Citation bot 1, ClueBot NG, Coinmanj, CommonsDelinker, DBigXray, Damien Linnane, Danim, Double sharp, El C, Elfrig, Extra999, Ezequiel828, Fastily, Fgnievinski, Finnusertop, Flyer22 Reborn, Gallina3795, Geoffrey.landis, Hameltion, Harlock81, Hbent, Headbomb, Huldragur, Johannes Animosus, JO4n, Jfmantis, Jodosma, JorisvS, Jusdafax, KAP03, Kogge, Kwamikagami, LahmacunKebab, KylieTastic, LedgendGamer, LennyDaBoss, MONGO, Materialscientist, Mygerardromance, NawlinWiki, Negative24, Nono64, Northernmac, Ohms law, Onel5969, Oshwah, Panda4999, Patient Zero, PohranicniStraze, Reaper Eternal, Reatlas, Rich Farmbrough, Rjwilmsi, Ruslik0, Sandstein8, Serendipodous, Serols, Shirulashem, Smith609, Spencerclift21, Stang, Steel1943, Telekenesis, The wub, Tom.Reding, Treehill, Tycho Magnetic Anomaly-1, Uncle Milty, WadeSimMiser, WikiPuppies, Wizardman, 97 anonymous edits 31

Climate of Uranus *Source:* https://en.wikipedia.org/w/index.php?oldid=851102764 *License:* Creative Commons Attribution-Share Alike 3.0 *Contributors:* Anupam, BilCat, Citation bot 1, ClueBot NG, Dewritech, Ego White Tray, Enterprisey, Epicgenius, FetchcommsAWB, Gene93k, Headbomb, Holdoffhunger, Jcpag2012, John of Reading, JorisvS, Lamb699, Line 8 the Pink, MartinZ, Nakon, Newyork1501, Novangelis, Obsidianspider, Pip2andahalf, Plastikspork, R'n'B, Reyk, Rich Farmbrough, Ricky81682, Rjwilmsi, RomanSpa, Ronhjones, Roux, Ruslik0, Saros136, Serendipodous, SkarmcA, The Thing That Should Not Be, Thegreatdr, Thordragon99, TommyBoy, Treehill, Tycho Magnetic Anomaly-1, UAwiki, Wavelength, West.andrew.g, Wikipelli, 61 anonymous edits 47

Formation and evolution of the Solar System *Source:* https://en.wikipedia.org/w/index.php?oldid=856485525 *License:* Creative Commons Attribution-Share Alike 3.0 *Contributors:* Agmartin, Arnusna, Antoni Barau, Arbnos, Ashill, Astronaut Willis, Audin, Aveh8, BatteryIncluded, Bear-rings, Bender235, Benfxmth, Besselfunctions, BirdValiant, Blackninga3216, BlakePlays, Bom319, CES1596, CLCStudent, CanadianLinuxUser, ChamithN, Charles Edwin Shipp, ClueBot NG, CyanoTex, Cyrus noto3at bulaga, DVdm, DatGuy, David.moreno72, Davros69999, Dawnseeker2000, Dbachmann, Dcirovic, Delphan Gruss EVI, DemocraticLuntz, Dodshe, Dolkungbighead, Download, DrKay, DrStrauss, Drbogdan, Drahkebmoon, Ehrenkater, El C, Eric Kvaalen, Escape Orbit, Excirial, FT2, Fama Clamosa, Fanta206, FlightTime, Friginator, Frinthruit, Gakrivas, Gap9551, GeneralizationsAreBad, GiantSnowman, Gilliam, Gilo1969, Hameltion, Hdjensofjfnen, Headbomb, Hillbillyholiday, Houjou, I dream of horses, Intel4004, Ira Leviton, Iridescent, IronGargoyle, J 1982, JFG, JaconaFrere, Jarble, Jcpag2012, Jim1138, Jmencisom, Joe Joe S.34567, John, JohnGolt, Jon Kolbert, JorisvS, Josve05a, Jwratner1, Kbh3rd, Kianu1234567891234679, Kolbasz, Kozmosis, LA1 FAL, Lawrence5641, Lchappell, Lugia2453, Mark-Sewath, Materialscientist, Mchcopl, Messier8, MusikAnimal, NSH002, Newone, Niceguyedc, Niseto, Orenburg1, Oshwah, Oxcross, Pfalzner, PlanetUser, Quenhitran, RA0808, RP88, Reatlas, Red Planet X (Hercolubus), Reddi, Rmosler2100, Rod57, Rothorpe, Ruslik0, Serendipodous, Serols, Shellwood, Sir Cumference, Sithlord Timmy, Slightsmile, Space Infinite, Steven Corder, Tetra quark, Tfr000, The Herald, ToBeFree, Tom.Reding, Treisijs, Urhixidur, V620 Cephei, Vgy7ujm, Vsmith, WikiProffy2122, Wtmitchell, Xinxinliu, ZaperaWiki44, Zezen, 174 anonymous edits 59

Moons of Uranus *Source:* https://en.wikipedia.org/w/index.php?oldid=851563538 *License:* Creative Commons Attribution-Share Alike 3.0 *Contributors:* 22hutchk, AbigailAbernathy, Ahonc, Akendall, Alcazar84, Amaury, Anomie, Anti-Quasar, Asturianu, BatteryIncluded, Bellerophon5685, BeywheelzLetItRip, BilCat, Bollyjeff, Br'er Rabbit, Byteflush, CielProfond, Citation bot 1, ClueBot NG, CommonsDelinker, Crystallizedcarbon, Cscslklk, DBigXray, Dcirovic, Denniss, Discospinster, Double sharp, El C, Fjörgynn, For i in range, Gallina3795, Gilderien, Ginsuloft, GoShow, GoingBatty, Graphium, Guoguo12, HMSLavender, Hans Dunkelberg, Headbomb, J 'mach' wust, J36miles, Jarodalien, Jjjjlolo, John, John of Reading, Jolielegal, Jonesey95, JorisvS, Jusdafax, Kheider, Kogge, Kwamikagami, KylieTastic, LahmacunKebab, Lithoicorn-138, Macauslan1984, Maczkopeti, Magicks, night94, Materialscientist, MelbourneStar, Nardog, Newone, Njardarlogar, Noneofyourbusiness, Ph0t0phobic, Pinethicket, Plastikspork, Pluto and Beyond, Razor2988, Reatlas, Richard3120, Rjwilmsi, Ruslik0, Ruslikrford1960, Rod57, Rother, Rothorpe, Ruslik0, Serendipodous, Serendipodous, Shellwood, Sizeofint, Solar-Wind, Spencerclift21, Ss112, Steel1943, StewartIM, TJH2018, Tamfang, Thbotch, Teles, The Thing That Should Not Be, Thincat, To wkepedia, Tom.Reding, Tomeasy, Tony Mach, Tycho Magnetic Anomaly-1, Violettsureme, Wikipelli, Xession, ﻻﺩﺮﻳ, 154 anonymous edits 85

Rings of Uranus *Source:* https://en.wikipedia.org/w/index.php?oldid=852742410 *License:* Creative Commons Attribution-Share Alike 3.0 *Contributors:* Acetotyce, Acroterion, Alansohn, Allens, Anonymous Dissident, Anton Gutsunaev, Art LaPella, Artman40, Atalkth, Azylber, Beat 768, Beniamino, Binky 2819, Bluebird207, Bongwarrior, Braini, CAPTAIN RAJU, Cashler, Cenarium, Citation bot 1, ClueBot NG, CommonsDelinker, Cyfal, Cédric Boissière, Dabomb87, Damicatz, Double sharp, DrKay, East718, Eeekster, Enggenchaz, Excirial, FKmailliW, Firsfron, Fjörgynn, Fotaun, Front Disco, Frosty, Galoubet, Gary, GateKeeper, Gene Nygaard, Giftlite, Gilliam, GrahamHardy, Greg L, GregorB, Guest9999, IRP, IronGargoyle, J 1982, J.delanoy, Jack Frost, Jacq9, Jhugh95, John, John of Reading, Johnmc, Julia W, KH-1, KTo288, Katydidit, Keith Edkins, Kethrus, Kheider, Korvin2050, Kwamikagami, Lasunncty, LilHelpa, LoneJedi1969, Los688, Lugia2453, MIDI, Mattisse, Mgiganteus1, Nakon, Naq, Nihilus, Ninney, Nouesrior, Novangelis, Openbookexam, Oshwah, Piledhigheranddeeper, Pizzaboys123, Plastikspork, Ququ, RA0808, Rab V, Ramisses, Reatlas, Rfassbind, Rich Farmbrough, Rjwilmsi, RomanSpa, Ruslik0, Ruby.red.roses, Ruslik0, Serendipodous, Signedzzz, Simplexity22, Spencerclift21, Steel1943, Stepheng3, T3dkjn89q00vl02Cxp1kqs3x7, The Roman Candle, TheCheese33, Tom.Reding, Tomerha91, Tomruen, Tony1, Trappist the monk, Treisijs, Twang, Tycho Magnetic Anomaly-1, Uncle Milty, Username Needed, Vengo-plus, Versus22, Wikiborg4711, Wikipelli, WolfmanSF, Wtshymanski, XLerate, ﻻﺏ, 111 anonymous edits 99

Exploration of Uranus *Source:* https://en.wikipedia.org/w/index.php?oldid=855966348 *License:* Creative Commons Attribution-Share Alike 3.0 *Contributors:* 0xDangit, Adavidb, AndrewHowse, Arielco, Astor14, BatteryIncluded, Bkell, ChiZeroOne, Chickyfuzz14, ChristinaGilbert, ClueBot NG, Cmglee, Colonies Chris, CommonsDelinker, CompuHacker, Crystal whacker, Czolgolz, DPBT1, Double sharp, Eannonder, Epbr123, Eyreland, Feces98, Fotaun, Frankturvl, Gilliam, Gits (Neo), Grafen, Groovykramer, Heureusementici, Imrek, Isambard Kingdom, JFG, Jeremyb-phone, Jschnur, K6ka, KylieTastic, LakesideMiners, LumberRift, MACGeeker, Marasama, Michaelmas1957, Mlm42, Name Omitted, NatureA16, Neumannk, Pious7, Pluma, Pmcray, Potterfreak0728, Primergrey, Qurq, Realkyhick, Reatlas, Rjwilmsi, Ruslik0, SSR2000, Sanfranman59, Saurusaurus, Semmendinger, Shellwood, Shexxehbeast2, Skizzik, Surajt88, Sus scrofa, Sylfred1977, Thbotch, TheLizardMan, Tom.Reding, Topatientlyexplain, Ushau97, VQuakr, Vacation9, Vipinhari, VoABot II, WereSpielChequers, Wjfox2005, Woohookitty, Wtmitchell, XavierGreen, Xession, 87 anonymous edits 117

Image Sources, Licenses and Contributors

The sources listed for each image provide more detailed licensing information including the copyright status, the copyright owner, and the license conditions.

Image *Source:* https://en.wikipedia.org/w/index.php?title=File:Padlock-silver.svg *Contributors:* AzaToth, BotMultichill, BotMultichillT, Gurch, Jarekt, Kallerna, Multichill, Perhelion, Rd232, Riana, Sarang, Siebrand, Steinsplitter, 4 anonymous edits 1
Image *Source:* https://en.wikipedia.org/w/index.php?title=File:Uranus2.jpg *License:* Public Domain *Contributors:* NASA/JPL-Caltech 1
Image *Source:* https://en.wikipedia.org/w/index.php?title=File:Loudspeaker.svg *License:* Public Domain *Contributors:* User:Dbenbenn, User:Optimager, User:Tsca, User:Dbenbenn, User:Optimager, User:Tsca, User:Dbenbenn, User:Optimager, User:Tsca 1
Figure 1 *Source:* https://en.wikipedia.org/w/index.php?title=File:William_Herschel01.jpg *Contributors:* Andrew Gray, ArtMechanic, Boo-Boo Baroo, BotMultichill, BotMultichillT, Dcoetzee, Deadstar, EMStephens, Ecummenic, Jochen Burghardt, Laura1822, Leyo, Madmedea, Materialscientist, Peppe83～commonswiki, Victuallers, WeHaKa, Yann, 霧木諒二 5
Figure 2 *Source:* https://en.wikipedia.org/w/index.php?title=File:HerschelTelescope.jpg *License:* Public Domain *Contributors:* Mike Young ... 5
Figure 3 *Source:* https://en.wikipedia.org/w/index.php?title=File:Uranusandrings.jpg *License:* Public domain *Contributors:* Hubble Space Telescope - NASA Marshall Space Flight Center 9
Figure 4 *Source:* https://en.wikipedia.org/w/index.php?title=File:Uranus_orientation_1985-2030.gif *Contributors:* User:Tomruen 10
Figure 5 *Source:* https://en.wikipedia.org/w/index.php?title=File:Uranus,_Earth_size_comparison_2.jpg *License:* Public Domain *Contributors:* User:Jcpag2012 12
Figure 6 *Source:* https://en.wikipedia.org/w/index.php?title=File:Uranus-intern-en.png *License:* Public Domain *Contributors:* Uranus-intern-de.png: FrancescoA derivative work: WolfmanSF (talk) 12
Figure 7 *Source:* https://en.wikipedia.org/w/index.php?title=File:Alien_aurorae_on_Uranus.jpg *Contributors:* Huntster, Jmencisom 16
Figure 8 *Source:* https://en.wikipedia.org/w/index.php?title=File:Tropospheric_profile_Uranus_new.svg *License:* Creative Commons Attribution-Sharealike 3.0 *Contributors:* Ruslik0 17
Figure 9 *Source:* https://en.wikipedia.org/w/index.php?title=File:Uranian_wind_speeds.png *License:* GNU Free Documentation License *Contributors:* Apocheir, BotMultichill, Cherubino, File Upload Bot (Magnus Manske), Harlock81, MGA73bot2, OgreBot 2, Sarang, Szczureq, Velma, 1 anonymous edits 18
Figure 10 *Source:* https://en.wikipedia.org/w/index.php?title=File:Uranian_Magnetic_field.gif *License:* Public Domain *Contributors:* Auntof6, El., Geek3, Inductiveload, McZusatz, Mewtow, Nanite, Shizhao, Stassats, Yarl 19
Figure 11 *Source:* https://en.wikipedia.org/w/index.php?title=File:Uranuscolour.png *License:* Public Domain *Contributors:* Deuar～commonswiki, PlanetUser, Remember the dot, Ruslik0, UV, 2 anonymous edits 20
Figure 12 *Source:* https://en.wikipedia.org/w/index.php?title=File:Uranus_Dark_spot.jpg *License:* Public domain *Contributors:* BotMultichill, Daniel 1992, File Upload Bot (Magnus Manske), Jaan513, Jcpag2012, OgreBot 2, Ruslik0, 2 anonymous edits 21
Figure 13 *Source:* https://en.wikipedia.org/w/index.php?title=File:Uranus_clouds.jpg *License:* Public domain *Contributors:* NASA, ESA, and M. Showalter (SETI Institute) 23
Figure 14 *Source:* https://en.wikipedia.org/w/index.php?title=File:Uranian_moon_montage.jpg *License:* Public Domain *Contributors:* NASA/JPL 25
Figure 15 *Source:* https://en.wikipedia.org/w/index.php?title=File:ESO_-_Uranus_(by).jpg *Contributors:* European Southern Observatory 25
Figure 16 *Source:* https://en.wikipedia.org/w/index.php?title=File:Uranus_rings_discovery.gif *License:* Creative Commons Attribution-Sharealike 3.0 *Contributors:* User:Orion 8 27
Figure 17 *Source:* https://en.wikipedia.org/w/index.php?title=File:Uranian_rings_scheme.png *License:* Public Domain *Contributors:* Ruslik0 . 27
Figure 18 *Source:* https://en.wikipedia.org/w/index.php?title=File:Uranuslight.jpg *License:* Public Domain *Contributors:* NASA 28
Figure 19 *Source:* https://en.wikipedia.org/w/index.php?title=File:Uranus_Final_Image.jpg *License:* Public Domain *Contributors:* NASA 29
Image *Source:* https://en.wikipedia.org/w/index.php?title=File:Uranus's_astrological_symbol.svg *License:* Public Domain *Contributors:* Lexicon 29
Image *Source:* https://en.wikipedia.org/w/index.php?title=File:Wikiquote-logo.svg *License:* Public Domain *Contributors:* Rei-artur 30
Image *Source:* https://en.wikipedia.org/w/index.php?title=File:Cscr-featured.svg *License:* GNU Lesser General Public License *Contributors:* Anomie 30
Image *Source:* https://en.wikipedia.org/w/index.php?title=File:Symbol_support_vote.svg *License:* Public Domain *Contributors:* Anomie, Fastily, Jo-Jo Eumerus 31
Figure 20 *Source:* https://en.wikipedia.org/w/index.php?title=File:Uranus2.jpg *License:* Public Domain *Contributors:* NASA/JPL-Caltech 32
Figure 21 *Source:* https://en.wikipedia.org/w/index.php?title=File:Tropospheric_profile_Uranus_new.svg *License:* Creative Commons Attribution-Sharealike 3.0 *Contributors:* Ruslik0 35
Figure 22 *Source:* https://en.wikipedia.org/w/index.php?title=File:Uranian_stratosphere.png *License:* GNU Free Documentation License *Contributors:* User:SreeBot 37
Figure 23 *Source:* https://en.wikipedia.org/w/index.php?title=File:Uranian_wind_speeds.png *License:* GNU Free Documentation License *Contributors:* Apocheir, BotMultichill, Cherubino, File Upload Bot (Magnus Manske), Harlock81, MGA73bot2, OgreBot 2, Sarang, Szczureq, Velma, 1 anonymous edits 40
Image *Source:* https://en.wikipedia.org/w/index.php?title=File:Lock-green.svg *License:* Creative Commons Zero *Contributors:* User:Trappist the monk 44
Image *Source:* https://en.wikipedia.org/w/index.php?title=File:Commons-logo.svg *License:* logo *Contributors:* Anomie, Callanecc, CambridgeBayWeather, Jo-Jo Eumerus, RHaworth 45
Figure 24 *Source:* https://en.wikipedia.org/w/index.php?title=File:Uranuscolour.png *License:* Public Domain *Contributors:* Deuar～commonswiki, PlanetUser, Remember the dot, Ruslik0, UV, 2 anonymous edits 48
Figure 25 *Source:* https://en.wikipedia.org/w/index.php?title=File:Uranus_clouds.jpg *License:* Public domain *Contributors:* NASA, ESA, and M. Showalter (SETI Institute) 48
Figure 26 *Source:* https://en.wikipedia.org/w/index.php?title=File:Uranus_Dark_spot.jpg *License:* Public domain *Contributors:* BotMultichill, Daniel 1992, File Upload Bot (Magnus Manske), Jaan513, Jcpag2012, OgreBot 2, Ruslik0, 2 anonymous edits 50
Figure 27 *Source:* https://en.wikipedia.org/w/index.php?title=File:Uranian_wind_speeds.png *License:* GNU Free Documentation License *Contributors:* Apocheir, BotMultichill, Cherubino, File Upload Bot (Magnus Manske), Harlock81, MGA73bot2, OgreBot 2, Sarang, Szczureq, Velma, 1 anonymous edits 51
Figure 28 *Source:* https://en.wikipedia.org/w/index.php?title=File:Seasonal_change_on_Uranus.jpg *License:* Public domain *Contributors:* NASA, ESA and M. Showalter (SETI Institute) 52
Figure 29 *Source:* https://en.wikipedia.org/w/index.php?title=File:Uranus_Seasonal_variability_v2.png *License:* Creative Commons Attribution-Sharealike 3.0 *Contributors:* Ruslik0 53
Figure 30 *Source:* https://en.wikipedia.org/w/index.php?title=File:Uranusandrings.jpg *License:* Public domain *Contributors:* Hubble Space Telescope - NASA Marshall Space Flight Center 55
Figure 31 *Source:* https://en.wikipedia.org/w/index.php?title=File:Uranus_depart.jpg *License:* Public Domain *Contributors:* NASA 56
Figure 32 *Source:* https://en.wikipedia.org/w/index.php?title=File:Protoplanetary-disk.jpg *License:* Public Domain *Contributors:* NASA 60
Figure 33 *Source:* https://en.wikipedia.org/w/index.php?title=File:Pierre-Simon_Laplace.jpg *License:* Public Domain *Contributors:* Ashill, Ecummenic, Elcobbola, Gene.arboit, Jimmy44, Laura1822, Leyo, Mattes, NicoScribe, Olivier, 霧木諒二 , 2 anonymous edits 61
Figure 34 *Source:* https://en.wikipedia.org/w/index.php?title=File:M42proplyds.jpg *License:* Public domain *Contributors:* C.R. O'Dell/Rice University; NASA 62
Figure 35 *Source:* https://en.wikipedia.org/w/index.php?title=File:Solarnebula.jpg *License:* Public Domain *Contributors:* William K. Hartmann, Planetary Science Institute, Tucson, Arizona UNIQ-ref-0-cf064b24719039b5-QINU 64
Figure 36 *Source:* https://en.wikipedia.org/w/index.php?title=File:Artist's_concept_of_collision_at_HD_172555.jpg *License:* Public Domain *Contributors:* NASA/JPL-Caltech 66
Figure 37 *Source:* https://en.wikipedia.org/w/index.php?title=File:Lhborbits.png *License:* Creative Commons Attribution-Sharealike 3.0,2.5,2.0,1.0 *Contributors:* en:User:AstroMark 68
Image *Source:* https://en.wikipedia.org/w/index.php?title=File:Interactive_icon.svg *Contributors:* User:Evolution and evolvability 71
Figure 38 *Source:* https://en.wikipedia.org/w/index.php?title=File:Barringer_Crater_aerial_photo_by_USGS.jpg *License:* Public Domain *Contributors:* USGS/D. Roddy 71
Figure 39 *Source:* https://en.wikipedia.org/w/index.php?title=File:Voyager_2_Neptune_and_Triton.jpg *License:* Public Domain *Contributors:* NASA / Jet Propulsion Lab 75

134

License

Index